T0093903

Birkhäuser

Oberwolfach Seminars

Volume 52

The workshops organized by the *Mathematisches Forschungsinstitut Oberwolfach* are intended to introduce students and young mathematicians to current fields of research. By means of these well-organized seminars, also scientists from other fields will be introduced to new mathematical ideas. The publication of these workshops in the series *Oberwolfach Seminars* (formerly *DMV seminar*) makes the material available to an even larger audience.

Renzo Cavalieri • Hannah Markwig •
Dhruv Ranganathan

Tropical and Logarithmic Methods in Enumerative Geometry

 Birkhäuser

Renzo Cavalieri
Department of Mathematics
Colorado State University
Fort Collins, CO, USA

Hannah Markwig
Department of Mathematics
University of Tübingen
Tübingen, Baden-Württemberg, Germany

Dhruv Ranganathan
Department of Pure Mathematics
and Mathematical Statistics
University of Cambridge
Cambridge, UK

ISSN 1661-237X   ISSN 2296-5041 (electronic)
Oberwolfach Seminars
ISBN 978-3-031-39400-3   ISBN 978-3-031-39401-0 (eBook)
https://doi.org/10.1007/978-3-031-39401-0

This book is published under the imprint Birkhäuser, www.birkhauser-science.com by the registered company Springer Nature Switzerland AG
The registered company address is: Gewerbestrasse 11, 6330 Cham, Switzerland

Paper in this product is recyclable.

Preface

This book is based on the lectures given at the Oberwolfach Seminar *Tropical Curves, Logarithmic Structures, and Enumerative Geometry*, held in Fall 2021.

How many circles are tangent to three given circles in the plane? This question, already asked by the ancient Greek geometer Apollonius, is an example of a problem in *enumerative geometry*. In enumerative geometry, we fix a class of geometric objects and some conditions that we require the objects to satisfy. The conditions are chosen in such a way that indeed only a finite number of objects satisfy them. Then we count how many of the objects satisfy the conditions. Many such questions are easy to pose, albeit very hard to answer—a fact that keeps attracting interest to this beautiful area of mathematics to this day. Schubert introduced many interesting ideas for solving enumerative problems in the nineteenth century and Hilbert posed as the 15th problem in his famous list of mathematical problems presented at the International Congress of Mathematics in Paris in 1900 constructing a rigorous foundation for Schubert's counting calculus. Enumerative geometry further developed into the study of Gromov-Witten theory at the end of the twentieth century, inspired by string theory and mathematical physics.

Tropical geometry is a modern area which allows an exchange of methods between algebraic geometry and combinatorics. It can be viewed as algebraic geometry over the tropical semifield with the operations max and $+$. Through a degeneration process called *tropicalization*, an algebraic variety is turned into a *tropical variety*. The latter is a polyhedral complex satisfying certain conditions. Tropical geometry is a flourishing area with fruitful connections to many other areas within mathematics, such as symplectic geometry, arithmetic geometry, mathematical physics and optimization, but also to areas in fields of application of mathematics such as economy, machine learning and computational biology. The foundational ideas appeared in different forms in the 1970s and 1980s, e.g. as Viro's patchworking method, Gelfand-Kapranov-Zelevinsky's amoebas and Maslov's dequantization, but also through work of Bergman, as well as Bieri and Groves. Only since the late 1990s has an effort been made to systematically exploit connections to questions in algebraic geometry, prominently represented in work of Mikhalkin and Sturmfels. A shining example of the successes of these methods have been to questions in enumerative geometry.

In 2002, following suggestions of Kontsevich, Mikhalkin initiated the use of tropical methods in enumerative geometry by proving the celebrated *Mikhalkin correspondence theorem* [Mik05]. The latter states that we can determine the numbers N_d of complex rational plane curves through $3d - 1$ points in general position by counting the appropriate *tropical curves*. Such *tropical plane curve counts* will be discussed in detail in Part II of this book. Many other correspondence theorems stating the equality of certain enumerative numbers and their tropical counterparts followed this original one. Among them are correspondence theorems for *Hurwitz numbers*, which count numbers of covers of Riemann surfaces satisfying fixed ramification data. These will be discussed in detail in Part II of this book.

Correspondence theorems paved the way for numerous interesting results in enumerative geometry making use of the analogous tropical counts: for example, an efficient recursive algorithm for computing Welschinger invariants (which can be thought of as real analogues of the plane curve counts discussed above) was developed by Itenberg, Kharlamov and Shustin using the Caporaso-Harris technique (see Chapter 10, Section 10.2). Also, several structural properties of double Hurwitz numbers, including statements about piecewise polynomiality and wall-crossing formulas, have been proved using tropical methods.

In recent years, the mathematical community came to realize that the degeneration process which we refer to as tropicalization can most efficiently be described in terms of *logarithmic geometry*, which is the main focus of Part I of this book. In enumerative geometry, both in algebraic and in tropical geometry, the curves satisfying the desired conditions can often be phrased in terms of a zero-dimensional subspace of a moduli space parametrizing certain curves or maps. Log geometry allows to put correspondences on the level of moduli spaces. That is, we can tropicalize (i.e. degenerate) a whole moduli space of interest and obtain a space that parametrizes the analogous tropical objects. Requiring a condition amounts to picking a subspace of the moduli space, and tropicalizing such a subspace we obtain a tropical subspace parametrizing the tropical objects that satisfy the analogous condition.

With this book, we present an overview of the use of logarithmic and tropical methods in enumerative geometry. In Part I, building on three lectures given by Dhruv Ranganathan, we lay the foundations for proofs of correspondence theorems using logarithmic geometry. The next two parts exploit correspondence theorems by using the tropical methods to gain new insights into enumerative problems. In Part II, we focus on Hurwitz theory counting maps of algebraic curves (resp. Riemann surfaces) to curves, i.e. to a 1-dimensional target. This part is based on lectures given by Renzo Cavalieri. In Part III, we focus on counts of tropical plane curves which can be viewed as images of maps of algebraic curves to a two-dimensional target. This part builds on lectures given by Hannah Markwig. Each part is accompanied by exercises which should be useful for every newcomer to the subject to gain practical experiences.

Acknowledgements

First and foremost we are grateful to Mathematisches Forschungsinstitut Oberwolfach for providing the opportunity and optimal environment to run a seminar on Tropical and Logarithmic Methods in Enumerative Geometry. R.C. has received support from Simons Collaboration Grant 420720 and NSF grant DMS-2100962. H.M. has received support from the SFB-TRR 195 "Symbolic Tools in Mathematics and their Application" of the German Research Foundation (DFG). D.R. has received support from EPSRC New Investigator Award EP/V051830/1.

Fort Collins, CO, USA   Renzo Cavalieri
Tübingen, Germany   Hannah Markwig
Cambridge, UK   Dhruv Ranganathan

Contents

About the Authors

Photo credit: Archives of the Mathematisches Forschungsinstitut Oberwolfach. With kind permission.

Renzo Cavalieri completed his PhD at University of Utah in 2005 under the direction of Aaron Bertram. He was a postdoc at University of Michigan under the mentorship of Bill Fulton for the following three years. In 2008, he became faculty at Colorado State University where he is currently a professor in the department of mathematics.

Hannah Markwig completed her PhD in 2006 at the University of Kaiserslautern in Germany, advised by Andreas Gathmann. She was a Postdoc at the Institute of Mathematics and its Applications in Minneapolis and at the University of Michigan in Ann Arbor, before she started a Juniorprofessorship at the University of Göttingen in 2008. In 2011, she moved to the University of the Saarland as a Professor, and in 2016 to the University of Tübingen.

Dhruv Ranganathan completed his PhD at Yale University in 2016 under the direction of Sam Payne. He was a CLE Moore Instructor at MIT and a member at the Institute for Advanced Study in 2017. Since 2019, he has been at the University of Cambridge, where he is currently a professor of mathematics.

The authors have worked together since 2013, on several projects related to the themes discussed in this book. They have taught several courses, including at MSRI, Stockholm, and of course in Oberwolfach. In addition to their shared love of mathematics, the authors enjoy hiking, cooking, music, and the life-altering card game known as "tichu".

Part I

Toric Geometry and Logarithmic Curve Counting

Introduction and Overview

1   Enumerative Geometry

The paradigm in modern enumerative geometry is to study a space X via the geometry and topology of natural moduli space $\mathcal{M}(X)$ associated to X. The moduli spaces of interest in Gromov–Witten theory are moduli spaces of curves C in X, or more precisely, of maps $f : C \to X$ from curves. It is natural to fix the topological data in the moduli problem: the arithmetic genus of C and the homology class of $f_*[C]$. One generally also needs to control the singularities of C, and specify a "stability condition" to obtain a reasonable algebro-geometric space. However, once this is done, we can extract numbers using the topology and geometry of $\mathcal{M}(X)$. These are called *enumerative invariants* or *curve counting invariants*.

Let us think about natural ways in which we can extract numbers. The simplest thing that one could do is calculate the topological Euler characteristic of $\mathcal{M}(X)$. While this will be a little coarser than what we want, a good example to keep in mind to start. If $\mathcal{M}(X)$ were, by some miracle, a compact complex manifold, then one could also use the *cohomology ring* of $\mathcal{M}(X)$. For example, if we choose a basis for the cohomology ring, then we can form polynomials in these cohomology classes of degree equal to the dimension of $\mathcal{M}(X)$. Then using the fact that we can integrate top degree cohomology classes on a compact manifold against the fundamental class to obtain a number.

In fact, we will essentially always use this method above to obtain numbers, though we should note that in modern Gromov–Witten theory, one rarely has such luck for $\mathcal{M}(X)$ to be a compact complex manifold. Nevertheless, a version of the above recipe can be done by using certain *specific* cohomology classes called tautological classes and a replacement for the fundamental class of $\mathcal{M}(X)$. The points we wish to illustrate here are, however, orthogonal to the virtual class and we will essentially ignore it from here on.

Even in the simplest instances, the computation of these enumerative invariants is extremely subtle and challenging. However, the invariants themselves are very rich, and their study is motivated from various different perspectives. The computation of enumerative invariants is usually carried out using a small number of techniques. The development of tropical techniques in the 2000s related these subtle invariants to a purely combinatorial problem—counting certain decorated graphs.

2   The Tropical Approach

Let us give a flavour of tropical curves and tropical curve counting, without worrying too much about precise definitions, which will come later. Two of the simplest statements in enumerative geometry are (1) that there exists a unique line through 2 distinct points in $\mathbb{P}^2$, and (2) that there exists a unique conic through 5 generically chosen points in $\mathbb{P}^2$. These two statements have perfect analogues in tropical geometry: there exists (1) a unique tropical line, whatever it might be, through 2 distinct points in $\mathbb{R}^2$, and (2) a unique tropical conic, whatever it might be, through 5 generically chosen points in $\mathbb{R}^2$. Figure 9.1 in Chap. 9 depicts these two curves.

The picture gives a general sense of what tropical curves look like, so let us make a note of these. First, tropical curves are composed of linear pieces—rays and line segments. Second, the vertices satisfy a "zero tension" or "balancing" condition: the sum of the outgoing edge directions at each vertex must be 0. The *degree* of the tropical curve is captured by the *asymptotic shape* of the curve, i.e. the picture one sees outside of any compact region. Putting these together, one can see that each of these curves is *rigid*. Precisely, there is no deformation of this picture that one can make while preserving the zero tension condition and the asymptotic shape.

In general, Mikhalkin's correspondence theorem tells you that enumerative invariants associated to curves passing through points in $\mathbb{P}^2$ can be computed in exactly this fashion—by counting tropical curves of a predictable shape, weighted by an appropriate multiplicity (which happened to be 1 in the cases above). You can learn about how to *actually* do this tropical counting in Part III.

The goal of these lectures is to understand the engine behind these tropical curve counting techniques, and how to think about the *correspondence theorem* that equates combinatorially weighted counts of tropical curves—certain piecewise linearly embedded graphs in $\mathbb{R}^n$—with counts of algebraic curves in projective space, and more generally in toric varieties. We will explain how to transfer the geometric problem into the combinatorial world. A number of paths exist to connect the two worlds, but one of them passes through *logarithmic geometry*, and specifically through logarithmic Gromov–Witten theory. This is the path that we will take in these lectures, but we will try to do so without really using any of the logarithmic technology. This is possible because we will focus on the most elementary of cases: counting curves of genus 0 in toric varieties. In this setting, all the structure that would be provided by this sophisticated theory of logarithmic stable maps, is provided by toric geometry, and by the theory of compactifications of subvarieties of tori.

The correspondence theorem here was proved in dimension 2 by Mikahlkin and in all dimensions by Nishinou–Siebert [Mik05, NS06], although we present a slightly different proof than those pursued in those papers.

3   Our Guiding Problem

Let us now outline the problem and the strategy, focusing on the question of counting curves passing through points in $\mathbb{P}^2$. We will leave out essentially all the technical details, and the rest of these lectures will be spent filling these in.

We are interested in curve counting problems, and to be concrete, let us consider the problem of counting genus 0, degree d curves in $\mathbb{P}^2$. To keep this concrete, let us consider maps

$$f : \mathbb{P}^1 \to \mathbb{P}^2,$$

of degree d, i.e. such that the f-preimage of a generic line consists of d points. We will consider two such maps to be isomorphic if they differ by a change of coordinates on the source.

There is a space $\mathcal{M}(\mathbb{P}^2, d)$ parameterizing these data, and an elementary parameter count, together with a soft deformation theory argument, shows that $\mathcal{M}(\mathbb{P}^2, d)$ is smooth and has dimension $N = 3d - 1$. In fact, we will later see an explicit parameterization of this space, so you can think about it in coordinates.

Let $p_1, \ldots p_N$ be a generic set of points in $\mathbb{P}^2$. The constraint that a point in $\mathcal{M}$ gives rise to a curve that passes through p_i is a codimension 1 constraint. Therefore, if we impose the condition that the curve passes through all points, we *expect* a zero dimensional set. We want to define N_d to be the size of this set. The first technical hurdle is defining this number rigorously in a way that it does not depend on the generic choice of points $p_1, \ldots, p_N$.

Once the number has been defined, we come to the point of actually computing it. We first consider a new space $\mathcal{M}_n(\mathbb{P}^2, d)$, where the points correspond to maps

$$f : (\mathbb{P}^1, q_1, \ldots, q_n) \to \mathbb{P}^2,$$

where $q_1, \ldots, q_n$ are distinct labelled points on the source $\mathbb{P}^1$. Given such a marked curve and a map, we can consider the following map

$$\text{ev} : \mathcal{M}_n(\mathbb{P}^2, d) \to (\mathbb{P}^2)^n.$$

which associates to each $[f]$ as above the tuple $(f(q_1), \ldots, f(q_n))$. The number we are interested in then is simply

$$N_d = \deg(\mathsf{ev}), \quad \text{when } n = N = 3d - 1.$$

We will later see that the space $\mathcal{M}_n(\mathbb{P}^2, d)$ is irreducible and the map ev is dominant. The degree of this map is therefore well-defined. Although we will solve the problem using tropical and logarithmic methods, you can find a more traditional approach to the problem in the excellent sources [FP97, KV07].

4   Compactifying the Problem

Although the above path is a very elementary definition, it does not immediately offer a method to calculate the numbers. One possible method of calculation is to replace the non-compact space $\mathcal{M}_n(\mathbb{P}^2, d)$ with a compactification, and then use cohomological methods to calculate the degree.

The space $\mathcal{M}_n(\mathbb{P}^2, d)$ can be compactified by allowing certain *degenerate* objects into the moduli problem. Once this is done, there is at least a theoretical path to calculation: find a collection of hypersurfaces in $(\mathbb{P}^2)^n$ that meet transversely at a single point, pull back these hypersurfaces to the proposed compactification, and perform an intersection theory calculation. Even this is far from straightforward to execute. We will use tropical geometry to carry this out in a more effective fashion.

Our strategy will be to pass to compactify the source of the morphism ev to

$$\mathsf{ev} : \overline{\mathcal{M}}_n^{\log}(\mathbb{P}^2, d) \to (\mathbb{P}^2)^n.$$

The compactification will be by the theory of *logarithmic stable maps*. This will be the central player. At first approximation, each point of $\overline{\mathcal{M}}_n^{\log}(\mathbb{P}^2, d)$ encodes (i) a map from a *potentially reducible* nodal curve to $\mathbb{P}^2$, and (ii) a compatible choice of a combinatorial datum, i.e. the *tropical curve*.

5   Calculation via Combinatorics

Setting aside the details of this central player, we can now finish the argument. The space $\overline{\mathcal{M}}_n^{\log}(\mathbb{P}^2, d)$ is very close to being a toric variety, and $(\mathbb{P}^2)^n$ is certainly a toric variety. By using combinatorial techniques, we will be able to find a further birational modification

$$\mathsf{ev}^\dagger : \overline{\mathcal{M}}_n^{\log}(\mathbb{P}^2, d)^\dagger \to \mathsf{Ev},$$

to ensure that the map is flat, and it is already proper. The flatness is guaranteed by the second technical ingredient, namely the *semistable reduction* theorems of Abramovich and Karu, and Adiprasito, Liu and Temkin [AK00, ALT18], which we use only the toric situation.

Calculating the degree of a flat and proper morphism is an accessible problem, and this is where the combinatorics enters. The source and target of $\mathsf{ev}^{\dagger}$ are both simple normal crossings pairs. Precisely, they are pairs (X, D) where X is smooth and D is a union of transversely intersecting smooth and irreducible divisors. They are therefore *stratified spaces*: the codimension k strata are points that lie in the intersection of k components of the natural boundary divisor, but not $k + 1$ of them.

Since the map is both flat and proper, we can compute the degree N_d by counting preimages of any point of the target, counted with their natural scheme theoretic multiplicity. The correspondence theorem now follows from choosing a 0-dimensional stratum of Ev and performing this calculation. We will be able to show that only 0-dimensional strata of the source map to it. These 0-dimensional strata can be understood as highly degenerate objects. While each logarithmic map gives rise to a tropical map, in these cases the tropical map determines the logarithmic map. These are indexed by tropical curves with predictable multiplicity. This leads to the correspondence theorem.

Geometry of Toric Varieties

The problem that will guide us throughout these notes is about compactifying moduli spaces. Typically, there are many different ways to compactify a variety and only some of them can be worked with in a geometrically meaningful way. One of the best developed theories of compactifications is that of *equivariant* compactifications of tori— toric varieties. Our strategy will be to use what is known in the toric setting to study our moduli spaces, which are not tori, via combinatorial techniques. In this section, we review the basic theory of toric varieties. The standard references here are [CLS11, Ful93].

1.1   Basics

Throughout these lectures, we will work over an algebraically closed field k of characteristic 0. Let T be an algebraic torus of dimension n; we'll use the notation $\mathbb{G}_m^n$ sometimes. The coordinate ring of T is the ring of Laurent polynomials, or more canonically can be viewed as the group ring M of a finitely generated abelian group. In other words, M is abstractly isomorphic to $\mathbb{Z}^n$ and we have

$$T = \mathrm{Spec}(k[M]).$$

Elements of M are monomial functions on T with coefficient 1; the isomorphism with $\mathbb{Z}^n$ records the exponents of a Laurent monomial.

Definition 1.1.1 A toric variety is a normal variety X equipped with a T-action

$$T \times X \to X$$

© The Author(s), under exclusive license to Springer Nature Switzerland AG 2023
R. Cavalieri et al., *Tropical and Logarithmic Methods in Enumerative Geometry*,
Oberwolfach Seminars 52, https://doi.org/10.1007/978-3-031-39401-0_1

equipped with a dense open orbit and an identification of this orbit with T in a manner that extends the group multiplication on T to the group action above.

The simplest examples are $\mathbb{A}^n$, $\mathbb{P}^n$, and products thereof. More generally, one can start with the torus T and partially compactify it in a way that respects the T action. Let us look at a combinatorial recipe to do this. Pick a finite collection of elements $\{u_1, \ldots, u_k\}$ in M and take their positive span. Call the result P; for simplicity, let us assume that the subgroup generated by P is M. Then define

$$U_P = \mathrm{Spec}(k[P]).$$

Certainly U_P contains T and one traces the definitions to verify that this is a toric variety although, strictly speaking, this need not be normal. This issue will go away in a moment when we work on the dual side. Alternatively, we can normalize to get a toric variety.

Example 1.1.2 If we take $M = \mathbb{Z}^2$ and examine the monoid generated by $(1,0)$, $(0,1)$, and $(0,-1)$. The monoid is abstractly isomorphic to $\mathbb{N} \times \mathbb{Z}$. The group ring is the ring $k[x, x^{-1}, y]$. The spectrum gives the toric variety U_P which is isomorphic to $\mathbb{A}^1 \times \mathbb{G}_m$.

One can generalize this to find monoids P such that the toric variety U_P is $\mathbb{A}^m \times \mathbb{G}_m^r$ for nonnegative integers m and r.

Example 1.1.3 In this example, we will construct a variety from equations inside affine space that come from a monoid. Equations will manifest as relations between generators of the monoid, so to get something nontrivial, we should take generators for the monoid that are linearly dependent, i.e. that have relations. For example, in $\mathbb{Z}^3$ take

$$(0,0,1), \quad (1,0,1), \quad (0,1,1), \quad (1,1,1).$$

The sum of the first and last vector is equal to the sum of the middle two. Using this, we find that $k[P]$ is isomorphic to $k[x, y, z, w]/(xw - yz)$; notice how the additive relation becomes a multiplicative binomial relation. The variety is the affine cone over $\mathbb{P}^1 \times \mathbb{P}^1$ in its standard embedding in $\mathbb{P}^3$.

One can produce more general toric varieties by gluing as in algebraic geometry, but rather than gluing along arbitrary open subsets, one only glues along open subsets that are *invariant under the torus action*. This is much better—there are only finitely many of these! It is more intuitive to do this via a dual picture.

1.2 The Dual Side

There is another lattice, typically denoted N defined by:

$$N = \text{Hom}(M, \mathbb{Z})$$

in abelian groups. We think of this as the lattice of algebraic homomorphisms

$$\mathbb{G}_m \to T.$$

Concretely, if M and therefore N are both identified with $\mathbb{Z}^n$, then the map determined by a tuple $(a_1, \ldots, a_n)$ sends λ in $\mathbb{G}_m$ to the tuple $(\lambda^{a_1}, \ldots, \lambda^{a_n})$. The pairing becomes the dot product.

We can interpret the pairing geometrically as well. Given a function χ in M and a map

$$\mathbb{G}_m \to T \xrightarrow{\chi} \mathbb{C}.$$

Either χ or $1/\chi$ will make sense at the missing 0 in $\mathbb{G}_m$, and therefore χ has a sensible order of vanishing, which is an integer. Precisely, expand χ as a Laurent series near 0. The pairing is the order of vanishing. (If this is not obvious to you, write down an example and you'll see a proof). A monoid P in M as above determines a dual cone by setting

$$\sigma_P = \{v \in N : \langle v, u \rangle \geq 0 \ \forall u \in P\}.$$

This notion of duality is actually a duality. You should check this. But it means that you can start from the N side and recover the monoid P that we already built.

A *cone* in $N_\mathbb{R}$ is the positive span of a finite set of lattice points in N. We will assume such a cone is always *strictly/strongly convex*, meaning it does not contain any lines. Each vector u in M defines a hyperplane. A *supporting hyperplane* is a hyperplane containing a cone in a half-space, i.e. the cone is entirely on the non-negative or non-positive side of the hyperplane. A *face* of a cone is the intersection of a cone with a supporting hyperplane.

Definition 1.2.1 A fan Σ is a collection of cones in $N_\mathbb{R}$, closed under taking faces, and such that the intersection of two cones is a face of each.

I will be a little loose and denote the toric variety associated to a cone σ also by U_σ. An unwinding of definitions shows that the toric variety of a face of a cone is an open subvariety of the toric variety of the larger cone. In fact, if u is the supporting hyperplane, then we just localize at χ^u.

The toric variety $X(\Sigma)$ is the glued scheme obtained from identifying the toric varieties U_σ across the opens determined by faces. One clean way to say this is that

$$X(\Sigma) = \text{colim}_{\sigma \in \Sigma} U_\sigma,$$

where the maps in the colimit diagram are given by open toric immersions.

It is a fundamental fact, due to Sumihiro, that every normal toric variety X comes from a fan Σ.

We record one more useful way of looking at the fan. Let X be a toric variety and T its torus as above. Given a point v in the lattice N, there is a one-parameter subgroup

$$\varphi_v : \mathbb{G}_m \to T \subset X.$$

We ask two questions: (i) Does the 1-parameter have a limit at 0? (ii) What is that limit?

The fan is precisely the structure that is obtained by partitioning the points of N for which the answer to (i) is yes, based on the answer to (ii). Said better, two points lie in the interior of the same cone if their limits at 0 are the same.

1.3   The Toric Dictionary

There is a basic dictionary between the geometry of toric varieties and the combinatoroics of their fans. Since this is not a lecture course on toric geometry, I will mention this rapidly, replacing all proofs by representative examples.

The following parts of the dictionary are about a single toric variety.

1. There is an order preserving bijection between affine torus invariant subvarieties and cones of Σ;
2. There is an order reversing bijection between torus orbits and cones of Σ;
3. A toric variety X is proper if every point of N lies in a cone of Σ; such a fan is called *complete*;
4. A toric variety X is smooth if every cone is generated by a subset of a $\mathbb{Z}$-basis for N.
5. A T-invariant principal divisor is a point of M, namely a linear function on N. Explicitly, evaluate the function at the generated of each ray and take the corresponding integer linear sum of divisors. A T-invariant Cartier divisor is a piecewise linear function on Σ, i.e. a continuous function on N that is linear on each cone in Σ.

Example 1.3.1 (The Dictionary for $\mathbb{P}^2$ and $\mathbb{A}^2$) Let us keep $\mathbb{A}^2$ and $\mathbb{P}^2$ in mind and go through the first three parts of the dictionary. The torus invariant opens in $\mathbb{A}^2$ are $\mathbb{A}^2$ itself, the opens $\mathbb{G}_m \times \mathbb{A}^1$ and $\mathbb{A}^1 \times \mathbb{G}_m$, and $\mathbb{G}_m^2$. Let us demonstrate how the affine $\mathbb{G}_m \times \mathbb{A}^1$ corresponds to a cone in the fan $\mathbb{R}^2_{\geq 0}$. If we take σ to be the positive y-axis, then the dual monoid P of positive functions on σ are the lattice points (a, b) such that $b > 0$. Then P is

$\mathbb{Z} \times \mathbb{N}$, which leads to the claim. The others are similar. The situation for $\mathbb{P}^2$ is analogous, with the invariant affine opens being 7 in number.

For the closed torus orbits in $\mathbb{P}^2$, these are obtained by choosing a subset of the homogeneous coordinates are declaring them to be 0 and forcing the remainder to be nonzero. There are 7 of these orbits in this case.

Now notice that $\mathbb{A}^2$ is not proper, and its fan is not complete. More precisely, the complement of the fan, i.e. $\mathbb{R}^2_{\geq 0}$ are those points (a, b) outside the first quadrant. Notice that if $v = (a, b)$ is a lattice point in this quadrant, there is a map

$$\varphi_v : \mathbb{G}_m \to \mathbb{A}^2$$

sending t to (t^a, t^b). Since at least one of these coordinates blows up, the limit does not exist. On the other hand, $\mathbb{P}^2$ is proper. In an exercise, you will calculate the different limits of the different 1-parameter subgroups and see the fan emerge from this perspective.

The classification by fans is functorial, so an equivariant morphism $f : X \to Y$ of toric varieties is given by a morphism $f_\star : \Sigma_X \to \Sigma_Y$ of fans, i.e. a map of lattices such that each cone gets sent to a cone. The following parts of the dictionary are about morphisms.

1. The morphism f is proper if and only if the preimage of the support of Σ_Y is contained in Σ_X.
2. A morphism is equidimensional if every cone of the source maps surjectively onto a cone of the target.
3. An equidimensional morphism has reduced fibers if and only if the image of the lattice of each cone in Σ_Y generates the corresponding lattice in Σ_X.

The properness is seen easily.

Example 1.3.2 Let Σ be a fan in $N_{\mathbb{R}}$. Let Δ be a subset of cones of Σ that also forms a fan. The inclusion $\Delta \to \Sigma$ defines a map of fans; it corresponds to an open immersion. Specifically, it is the inclusion of the complement of the union of torus orbits corresponding to the cones of Σ not in Δ. It is essentially never proper.

Example 1.3.3 A typical example of a proper morphism is induced by a *subdivision* or *refinement*. That is, given a fan Σ in $N_{\mathbb{R}}$, consider Δ to be a fan in the same space $N_{\mathbb{R}}$, equipped with a map $\Delta \to \Sigma$ that is a bijection on the supports. The map is certainly proper, and corresponds to a proper birational morphism, e.g. a blowup at a torus invariant subscheme.

Finally, there are cohomological invariants. You may not have seen this, and the ideas are more sophisticated but if you are interested in intersection theory you may enjoy this.

1. Every divisor—both Weil and Cartier—on a toric variety X is equivalent to a a linear combination of T-invariant divisors. Since the latter are in bijection with the rays of the fan Σ of X, there is a surjective map from the free abelian group on the rays of Σ to the Weil divisor class group.
2. Each character $\chi \in M$ gives rise to a principal T-invariant Weil divisor. In fact, there is a simple formula for this:

$$\mathrm{div}(\chi) = \sum_{\rho} \langle v_{\rho}, \chi \rangle D_{\rho}$$

   where ρ runs over the rays in Σ, and v_{ρ} is the primitive generator on the ray. Together with the previous comment, this describes the Weil divisor class group completely.
3. A *Cartier* divisor on X is described by a *piecewise linear function* on the fan Σ, namely a $\mathbb{R}$-valued function on $|\Sigma|$ that restricts to an element on M on each cone. The Cartier divisor associated to a piecewise linear function φ is again given by the formula above: add up D_{ρ}'s with the coefficient given by evaluating φ on v_{ρ}.
4. If X is smooth and projective, then the Chow ring (or cohomology ring) is generated by the T-invariant divisors. An explicit presentation is given as follows. Let Σ be the fan, let $\Sigma^{(1)}$ denote the set of rays, and if ρ_j is a ray we let v_j be the corresponding set of generators. The ring is given by

$$A^{\star}(X) = \mathbb{Z}[D_{\rho}]_{\rho \in \Sigma^{(1)}}/I$$

   where the ideal I contains relations
   (a) $D_{\alpha_1} \cdots D_{\alpha_k}$ when $\rho_{\alpha_1}, \ldots, \rho_{\alpha_k}$ do not form a cone of Σ,
   (b) $\sum_{\rho_i \in \Sigma^{(1)}} \langle v_i, u \rangle$ for each fixed $u \in M$.
   The first relation accounts for the geometric situation where a set of divisors do not meet, while the second keeps track of relations in the Picard group of X.
5. If one drops the second set of relations above, the resulting ring is much larger. In fact, it is a very interesting ring known as the *equivariant Chow ring of X*.
6. There is an interpretation of this ring as the ring of piecewise polynomial functions on Σ. A cone σ comes with a sensible notion of linear function. Precisely, σ is a subset of N and a linear function is any function given by evaluating a fixed u in M under the pairing. As a consequence, there is a well-defined notion of *polynomial function* given by sums of products of linear functions and constants. Let $|\Sigma|$ be the *support* of Σ, given by set of all points in N contained in some cone of Σ. A *piecewise polynomial function* is a function

$$\varphi : |\Sigma| \to \mathbb{R},$$

whose restriction to any σ in Σ is a polynomial function. The ring of piecewise polynomial functions on the fan Σ of a smooth toric variety is naturally isomorphic to the ring discussed in the point above (prove this!). If Σ is not smooth, then the ring $PP^\star(\Sigma)$ is still very interesting: it is isomorphic to the equivariant operational Chow cohomology of $X(\Sigma)$, by a result of Brion and Payne [Bri97, Pay06].

7. In general if X is toric and singular, there is a difference between Chow homology and Chow cohomology. The latter is a slightly subtle theory, however, the key thing is that piecewise polynomials produce things that act on Chow homology. The T-equivariant Chow cohomology of a toric variety X is the ring of piecewise polynomial functions on the fan Σ.

1.4   A Few Toric Exercises

We start with some basic exercises that will get you oriented with the toric dictionary.

1. Consider the fan with 4 rays, spanned by $(\pm 1, 0)$ and $(0, \pm 1)$; the two dimension cones are spanned by pairs of clockwise consecutive rays. Prove that the resulting toric variety is $\mathbb{P}^1 \times \mathbb{P}^1$.
2. The fan of $\mathbb{A}^2$ is a single 2-dimensional cone spanned by $(1, 0)$ and $(0, 1)$. Subdivide this cone by introducing the ray along $(1, 1)$, breaking the 2-dimensional cones into two. By using the gluing description of toric varieties obtained from fans, prove that the resulting toric variety is the blowup of $\mathbb{A}^2$ at a point.
3. Consider the cone σ spanned by $(1, 0)$ and $(1, 2)$ in $N_\mathbb{R}$. Write out the coordinate ring $\mathbb{C}[S_\sigma]$ as a quotient of a polynomial algebra. Conclude that this toric variety is singular.
4. Consider the cone σ in $\mathbb{R}^3$ given by the cone over a square between vertices $(0, 0, 1)$, $(0, 1, 1)$, $(1, 1, 1)$, and $(1, 0, 1)$. By using the toric dictionary, find a Weil divisor that is not Cartier.
5. Use the toric dictionary to calculate the Picard group of $\mathbb{P}^2$ and $\mathbb{P}^1 \times \mathbb{P}^1$, and verify that this gives the expected answer.
6. Calculate the Chow and cohomology rings of a product of projective spaces.
7. Calculate the Chow ring of the blowup of $\mathbb{P}^2$ at a torus fixed point.

We now examine some slightly more sophisticated ones.

1. Construct a cone σ of dimension 3 such that the associated toric variety has a singular point with tangent space dimension k for arbitrarily large k.
2. Let Σ be the fan of $\mathbb{P}^1$—it consists of two top dimensional cones: the positive and negative half ways starting at 0. Calculate the ring of piecewise polynomial functions on Σ. Quotient by the ideal generated by globally linear functions to deduce that this ring is isomorphic to the cohomology ring of $\mathbb{P}^1$.

3. Let $\Sigma' \to \Sigma$ be a proper morphism of fans in $N'_{\mathbb{R}}$ and $N_{\mathbb{R}}$ respectively, induced by an integer linear map $N' \to N$. Prove that there exist subdivisions $\Delta' \to \Sigma'$ and $\Delta \to \Sigma$, such that the map $N' \to N$ naturally induces a map of fans $\Delta' \to \Delta$, with the property that every cone of the source surjects onto a cone of the target.

4. Prove that a fiber product of toric varieties with equivariant morphisms is not toric. Does the category of fans admit fiber products?

Compactifying Subvarieties of Tori

2

In this chapter, we will develop some techniques for compactifying varieties in general. In general, the problem of compactifying moduli spaces is difficult and very rich. We will be lucky in these lectures that the moduli problem that we want to compactify has some addition structure—it is "very affine", i.e. it is a clsoed subvariety of an algebraic torus. This enables us to use some finer aspects of toric geometry to compactify these moduli spaces. The general techniques will be developed in this chapter, and applied to moduli spaces of interest in the next two chapters.

The main reference for this section is Tevelev's paper [Tev07] and the text by Maclagan–Sturmfels [MS15], with significant inspiration from Kapranov's work [Kap93].

2.1   Compactifying Subvarieties of Tori

The simplest way to compactify varieties is by taking closure in something already compact. However arbitrary closure operations usually lead to compactifications we don't understand. Closure operations in well-chosen toric varieties, however, are more controllable. Here is our starting point.

Question Let $Z \hookrightarrow T$ be a subvariety. What is a best toric variety X with dense torus T in order to compactify Z?

This is a vague question, but two reasonable properties to look for are (1) that the closure $\overline{Z}$ is actually compact, and (2) that the orbits of T in X meet the closure of Z nicely. For example, we can ask for the intersection of $\overline{Z}$ with an orbit of codimension k to be exactly codimension k in $\overline{Z}$?

There are two equivalent ways to phrase this.

© The Author(s), under exclusive license to Springer Nature Switzerland AG 2023
R. Cavalieri et al., *Tropical and Logarithmic Methods in Enumerative Geometry*,
Oberwolfach Seminars 52, https://doi.org/10.1007/978-3-031-39401-0_2

1. The multiplication map

$$\overline{Z} \times T \to X$$

is flat and surjective; in particular, every fiber is non-empty and of the same dimension.
2. The morphism of stacks

$$\overline{Z} \to [X/T]$$

is flat and surjective.

If this property holds, we call the closure $\overline{Z} \hookrightarrow X$ a *tropical compactification*.

These mean exactly the same thing. If you have not seen stacks before, this is a good way to stop being afraid of them. The second option suggests natural generalizations in logarithmic geometry, though we will not need this.

Geometrically, the surjectivity corresponds to the statement that $\overline{Z}$ meets every torus invariant stratum of X. The flatness is subtle but it implies equidimensionality. This implies that if we have $\overline{Z}$ in X and a torus orbit W in X that has codimension k, the intersection $\overline{Z} \cap W$ has codimension k in $\overline{Z}$. In particular, $\overline{Z}$ should miss torus orbits of codimension larger than its own dimension.

Another point to note is that the surjectivity is not so difficult to achieve: if you have found X which satisfies the flatness but not surjectivity, you can simply throw away strata until it is surjective.

Now let us see a few simple examples of what a tropical compactification can and cannot look like.

Example 2.1.1 Suppose T is $\mathbb{G}_m^2$ with coordinates x and y. Let Z be the subvariety $\mathbb{V}(x - y)$. If we take X to be $\mathbb{P}^1 \times \mathbb{P}^1$, the closure $\overline{Z}$ is not a tropical compactification. It does not meet all the strata for one, but more crucially, notice that we have a curve meeting a torus fixed points; but a point is codimension 2 in $\mathbb{P}^1 \times \mathbb{P}^1$ and should not meet this curve at all.

When we look at examples, it will be convenient to not have to worry about the subtler points concerning flatness. If $\overline{Z}$ is Cohen–Macaulay, for example smooth, then flatness is equivalent to the statement that $\overline{Z}$ meets the torus orbits in the expected dimension.

Example 2.1.2 The simplest example of a tropical compactification is to take T to again be $\mathbb{G}_m^2$ with coordinates x and y, and take Z to be $\mathbb{V}(x + y + 1)$. If we take X' to be $\mathbb{P}^2$ then the closure $\overline{Z}$ is given by $\mathbb{V}(X + Y + Z)$ in homogeneous coordinates. Then one easily checks that $\overline{Z}$ misses the points $(1 : 0 : 0)$, $(0 : 1 : 0)$, and $(0 : 0 : 1)$. If we take X to be X' minus these three points; this is a toric variety with 4 torus orbits. Now $\overline{Z}$ meets all torus orbits in the expected dimension. Since $\overline{Z}$ is $\mathbb{P}^1$, and in particular is smooth, so Cohen–Macualay, the remark above guarantees that $\overline{Z}$ is a tropical compactification in X.

Our main tool for constructing compactifications comes from the following beautiful theorem of Tevelev.

Theorem 2.1.3 *Fixing $Z \hookrightarrow T$, there exists a toric variety X such that the closure $\overline{Z}$ is a tropical compactification of Z. Furthermore, if $X' \to X$ is a proper and birational map, then the closure $\overline{Z}'$ in X' is also tropical. Finally, $\overline{Z}'$ is the scheme theoretic preimage of $\overline{Z}$.*

I will sketch the construction of this, because the argument is very beautiful.

Proof By replacing Z and T by quotients, we can reduce to the case where Z is not fixed by a subtorus. Let Y be any projective toric compactification of T and let $\widetilde{Z}$ be the closure. Consider the point $[\widetilde{Z}]$ in the Hilbert scheme $\mathrm{Hilb}(Y)$. There is an action T on this Hilbert scheme. Let X be the normalization of the closure of the orbit of $[Z]$ under the torus action. This is a compactification and will be the one we need. We just need to show that the multiplication map is flat.

We certainly have an equivariant morphism

$$X \to \mathrm{Hilb}(Y).$$

The Hilbert scheme comes with a universal structure

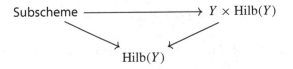

Pullback the universal subscheme to X via this morphism and denote the result by V; it is naturally a subscheme of $X \times \mathrm{Hilb}(Y)$ so we can sensibly restrict the result to $X \times T$. We will now identify this V with $\overline{Z} \times T$ and the natural flat map to X with the multiplucation map. Modulo these identifications, the claim follows.

As for the identifications, these are nice diagram chases. By definition, V can be described as

$$V = \{([W], w) \in X \times T : [W] \in X; \ w \in W \cap T\},$$

since a point on the universal family is a subscheme and a point on it. It is convenient to translate by w, and identify

$$V \cong \{([w^{-1}W], w) \in X \times T : [W] \in X; \ w \in W \cap T\}.$$

Now it suffices to show that $\overline{Z}$ is precisely the set of points in X whose corresponding subscheme contains the identity of the torus. This certainly contains Z but it's also closed, so $\overline{Z}$ must be a component of it. On the other hand, this set is irreducible, by flatness as you can easily check.

$\square$

One of the crucial properties that we gain from $\overline{Z} \to X$ being a tropical compactification is a type of stability under *further* birational modification. Precisely, if $X' \to X$ is a toric birational morphism, then the strict transform of Z is equal to its preimage under $X' \to X$.

By virtue of the "equidimensionality" aspect of the tropical compactification condition, it also follows that every codimension r boundary cycle in the toric variety X gives us such a boundary cycle in $\overline{Z}$. In particular, we have a large and natural supply of Weil and Cartier divisors on $\overline{Z}$ with combinatorial description. A very natural question is to measure how much of the Chow/homology groups these cycles generate. As we will see, in the moduli settings that we are interested in, the toric variety X sees the full Chow ring.

2.2   Tropicalization via Compactifications

Let $Z \subset T$ be a subvariety of a torus as in the previous section. Let X be a sufficiently refined toric variety so that $\overline{Z} \hookrightarrow X$ is tropical—we know such a thing exists. Define

$$\mathrm{trop}(Z) = |\Sigma_X| \subset N_\mathbb{R},$$

that is, the tropicalization of Z is the support of the fan of any tropical compactification.

Proposition 2.2.1 *The tropicalization defined above is independent of choice of X.*

Proof First note that any two compactifications can be dominated by a third. To see this, use the closure of the graph of a rational map between them. This reduces us to the case of comparing a tropicalization with respect to X with one with respect to X' a blowup of X. However, the flatness guarantees that $\overline{Z}' \subset X'$ is just the scheme theoretic preimage of $\overline{Z}$, so the result follows.

$\square$

The proof reveals a way to think about the tropicalization via toric geometry. Given a subvariety $Z \hookrightarrow T$, for each toric compactification X_Σ, we can ask which cones of Σ are dual to orbits that meet $\overline{Z}$. The union of the points in those cones are the tropicalization. This will depend on Σ but will eventually stabilize as we make Σ finer and finer.

Example 2.2.2 To drive the above point home, let's do an example. Given $Z \hookrightarrow T$ and an arbitrary compactification X, define the *fake tropicalization* to be the union of the cones of X which are dual to orbits that meet $\overline{Z}$. This fake tropicalization genuinely depends on X. Take Z to be the diagonal line in $\mathbb{G}_m^2$. Close it up in $\mathbb{P}^2$ first. The fake tropicalization is the union of the ray through $(-1, -1)$ and the positive quadrant. If we instead take the closure in the blowup of $\mathbb{P}^2$ at the point corresponding to the first quadrant, the fake (and genuine) tropicalization is just the diagonal line in $\mathbb{R}^2$.

Let us now compute a few more examples of tropicalizations.

Example 2.2.3 The simplest cases of tropicalization occur for hypersurfaces. Let $Z \subset T$ be a hypersurface defined by a Lauren polynomial $\sum a_u \chi^u$ where u is an element of the lattice M, and the a_u are nonzero complex numbers. The *Newton polytope* of Z is the convex hull of the lattice points u appearing in the sum above. The tropicalization of Z is, in this case, precisely the codimension 1 skeleton dual fan of this polytope. See [MS15] for a proof.

Example 2.2.4 Let $Z \subset \mathbb{P}^n$ be a generic linear subspace of dimension k. Since it is generic, Z will only meet the torus orbits in $\mathbb{P}^2$, which are of course just the coordinate subspaces, of expected dimension. It follows that trop(Z) is the union of cones in the fan of $\mathbb{P}^n$ of dimension k.

The story of tropicalizations for linear subspaces that are non-generic is very interesting—it leads directly to the theory of matroids. In order to connect this to what we have done, suppose we have a linear $Z \subset \mathbb{P}^n$ of dimension r. The appropriate Hilbert scheme arising in the *proof* of the existence of tropical compactifications is the Grassmannian of linear subspaces of dimension $r + 1$ in a vector space of dimension $n + 1$. The proof considered orbit closures of points in this Grassmannian. The toric varieties arising as orbit closures of the r-dimensional torus have natural polytopes, polarized by the Plücker embedding. They are subpolytopes of the hypersimplices $\Delta(r + 1, n + 1)$, and in fact are essentially the class of (realizable) matroid polytopes. See [MS15] for further details on this connection, as well as many more examples of tropicalization.

2.3 Tropicalizations via Valuations

The existence of tropical compactifications is a powerful theorem, the best you can hope for, but closures are typically useless operations unless you can get some handle on them. This is where tropical geometry, in the traditional sense, actually comes in. For this discussion, we set K to be the Puiseux series field, i.e.

$$K = \bigcup_n \mathbb{C}((t^{1/n})).$$

There is a natural valuation map on this field

$$K^\star \to \mathbb{Q}$$

recording the exponent of the leading term, see also Sect. 9.3 in Part III, Chap. 9.

Definition 2.3.1 The tropicalization of $Z \hookrightarrow T$ is defined as the closure of the image of $Z(K)$ under the "coordinatewise valuation" map

$$\text{trop} : T(K) \to N_{\mathbb{R}}$$

As an exercise, decode what exactly this means.

The two notions of tropicalization coincide, by foundational results in the subject. When we want to distinguish the definitions, we will use *valuative tropicalization* for the above definition and *geometric tropicalization* for the earlier one.

A basic result is the Bieri–Groves theorem for the valuative tropicalization.

Theorem 2.3.2 *The valuative tropicalization of $Z \hookrightarrow T$ is topological subspace of $N_{\mathbb{R}}$ of dimension equal to the algebraic dimension of Z, It is the underlying set of a fan.*

I will not provide a proof of this theorem; the best modern proofs involve a little bit of model theory and logic in the form of quantifier elimination. There are more algebraic and geometric proofs as well.

With regard to the comparison between the two tropicalizations, one direction is more or less straightforward.

Proposition 2.3.3 *The valuative tropicalization is contained in the geometric tropicalization.*

Proof I essentially leave this as an exercise; the basic point is to use the fact that if Σ is a fan and p is a K-point of T, then the limit of p exists if and only if the tropicalization of p is contained in Σ. This is the same logic that drives the equivalence between $X(\Sigma)$ being proper and Σ being complete. The result follows from applying the valuative criterion of properness.   □

In the reverse direction, one needs to construct a K-point with a given tropicalization. This inevitably involves the equations for Z. It can be done in various ways, for example using initial degenerations, or Newton's method, or the non-archimedean Nullstellensatz. I will not get into it here.

In particular, one thing becomes clear that was not obvious in our discussion so far.

Proposition 2.3.4 *Let Σ be a fan in $N_{\mathbb{R}}$ with toric variety X, and $Z \hookrightarrow T$ be a subvariety as above. The closure in X is proper if and only if the valuative tropicalization of Z is contained in the support of Σ.*

Depending on the context, the definition via valuations or via compactifications may be easier to compute.

2.4 Varying the Coefficients

In this setup, we have insisted that Z is defined over the ground field $\mathbb{C}$. As we have seen above, one definition of tropicalization is to look pass to a large non-archimedean extension and then look at the image of the K-points under the valuation map. It is natural to allow the variety Z to be defined over this field K, i.e. to allow the coefficients to vary.

Let us fix a copy of $\mathbb{C}[t] \subset K$, with the valuation on K restricting to order of vanishing at 0. In the simplest case, $Z \subset T$ is defined by an ideal I in $K[M]$, generated by polynomials f_i, whose coefficients lie in $\mathbb{C}[t, t^{-1}]$. In this case, we can think of the variety $W \subset T \times \mathbb{G}_m$ and apply the tropical compactification techniques above. We are left with $\overline{W} \subset Y$, where Y is a compactification of $T \times \mathbb{G}_m$. By projecting rationally onto the second factor, compactifying that factor by $\mathbb{P}^1$, resolving by toric blowups, and restricting to a neighborhood of 0 in $\mathbb{P}^1$, we are left with

$$\overline{W} \hookrightarrow \mathcal{Y} \to \mathrm{Spec}(R).$$

We have abused notation slightly by neglecting to decorate the restriction of $\overline{W}$. The general fiber of this map compactifies Z above. The variety $\mathcal{Y} \to \mathrm{Spec}(R)$ is the restriction to a formal neighborhood of the origin of a toric variety equipped with a map to $\mathbb{P}^1$.

Viewing it differently, given the subvariety Z in T with coefficients as above, we have found a compactification of T over $\mathrm{Spec}(K)$ and a degeneration $\mathcal{Y}$ of that compactification over $\mathrm{Spec}(R)$ such that the closure of Z is a flat family of subschemes inside $\mathcal{Y}$ and it has the flatness properties we discussed above.

In this more general context, one can define the *tropicalization* of Z as follows. We have a map of fans

$$\Sigma_{\mathcal{Y}} \to \mathbb{R}_{\geq 0}.$$

Then define trop(Z) to be the fiber over the point 1 of this map; it is a polyhedral complex.

The assumptions above, namely that the coefficients of the defining equations of Z lie in the Laurent ring $\mathbb{C}[t, t^{-1}]$, can be dropped. Let $Z \subset \mathbb{T}$ be defined by Laurent polynomials with coefficients in K. We have the following general form of the Bieri–Groves theorem.

Theorem 2.4.1 *The tropicalization of* $Z \hookrightarrow T$ *is topological subspace of* $N_{\mathbb{R}}$ *of dimension equal to the algebraic dimension of Z. It is the underlying set of a polyhedral complex.*

As before, the two notions of tropicalization—via compactification-degeneration and by images under valuation maps—coincide. The details may be found in Gubler's text [Gub13].

Points on the Riemann Sphere

3

We first use our machine to compactify the moduli space $\mathcal{M}_{0,n}$. This is a construction due to Kapranov, with its tropical context clarified by Tevelev [Tev07] and then by Gibney–Maclagan [GM10]. We start with

$$\mathcal{M}_{0,n} = \{(\mathbb{P}^1, p_1, \ldots, p_n) : p_i \text{ are distinct}\}/PGL.$$

In his work on Chow quotients of Grassmannians [Kap93], Kapranov teaches us that there are two ways to think about this:

1. **Abelian-by-Non-Abelian.** As a quotient of an open subset of $(\mathbb{P}^1)^n$ by $PGL(2)$.
2. **Non-Abelian-by-Abelian.** As a quotient of an open subset of $Gr(2, n)$ by the group $\mathbb{G}_m^{n-1}$.

These are the two sides of the "Gelfand–Macpherson correspondence". The perspectives are both useful, usually one for conceptual arguments and the other for calculations.

3.1   Putting $\mathcal{M}_{0,n}$ Into a Torus

Let us consider the second description of $\mathcal{M}_{0,n}$ above. Observe that if we have a generic line in $\mathbb{P}^{n-1}$, the intersection of that line with the n coordinate hyperplanes give rise to a configuration of n points on $\mathbb{P}^1$. Conversely, any such configuration of n points on $\mathbb{P}^1$ can be realized in this way.

The following two exercise spell out the details. Suppose we are given a 2-plane in $\mathbb{C}^n$. Equivalently we have a corresponding $\mathbb{P}^1$ in $\mathbb{P}^{n-1}$ embedded linearly. We say a line is *generic* if it misses all codimension 2 coordinate planes in $\mathbb{P}^{n-1}$.

© The Author(s), under exclusive license to Springer Nature Switzerland AG 2023
R. Cavalieri et al., *Tropical and Logarithmic Methods in Enumerative Geometry*,
Oberwolfach Seminars 52, https://doi.org/10.1007/978-3-031-39401-0_3

Exercise 3.1 Check that this notion of genericity is the same as the corresponding Plücker coordinates all being nonzero.

Given a generic line in $\mathbb{P}^{n-1}$, the intersection of the line with the n coordinate hyperplane gives n points on $\mathbb{P}^1$. The converse is a nice linear algebra exercise.

Exercise 3.2 Given n points on $\mathbb{P}^1$, with the first 3 at $0, 1, \infty$, show that this configuration can be realized from a generic line by hyperplane intersections as above. Prove that the set of such lines has a free and transitive action of $\mathbb{G}_m^n$.

Therefore, we can identify

$$\mathcal{M}_{0,n} = G(2,n)^{\circ}/\mathbb{G}_m^{n-1}$$

where T is the $n-1$-dimensional torus acting on the Grassmannian, and we have passed to the open subset where all the Plücker coordinates are nonzero. This torus is a subgroup of the ambient Plücker torus, so we have

$$\mathcal{M}_{0,n} = G(2,n)^{\circ}/\mathbb{G}_m^{n-1} \hookrightarrow \mathbb{G}_m^{\binom{n}{2}-1}/\mathbb{G}_m^{n-1}.$$

This lets us think of $\mathcal{M}_{0,n}$ as a subvariety of a torus, and gives access to the tools in the previous section.

We will use this torus to study $\mathcal{M}_{0,n}$ and its compactification by using tropical compactification techniques. The choice of torus may initially seem non-canonical—for example, why not take a bigger torus?

Proposition 3.1.1 *Let T' be an algebraic torus. Then every morphism*

$$\mathcal{M}_{0,n} \to T'$$

factors through the embedding

$$\mathcal{M}_{0,n} \hookrightarrow \mathbb{G}_m^{\binom{n}{2}-1}/\mathbb{G}_m^{n-1}.$$

via a homomorphism to T'.

Any variety X—with very minor hypotheses—has an *intrinsic* torus, namely the torus with character lattice $\mathcal{O}_X^{\star}/k^{\star}$. The space $\mathcal{M}_{0,n}$ is naturally a subset of $\mathbb{P}^{n-3}$ given by the complement of a collection of hyperplanes. In the case of complements of such hyperplanes like $\mathcal{M}_{0,n}$, it is straightforward to describe all the invertible functions.

Exercise 3.3 Calculate the rank of the intrinsic torus of $\mathcal{M}_{0,n}$ and describe functions corresponding to a basis for the lattice of this torus.

Hint: By the Abelian-by-Non-Abelian description, and using Möbius maps, view $\mathcal{M}_{0,n}$ is a subset of $(\mathbb{A}^1)^{n-3}$; the latter has trivial intrinsic torus. Now reason by studying what happens every time a hyperplane gets deleted.

The exercise above gives an explicit description of the functions that give the characters on the torus.

Exercise 3.4 Prove that the torus $\mathbb{G}_m^{\binom{n}{2}-1}/\mathbb{G}_m^{n-1}$ described above is the intrinsic torus of $\mathcal{M}_{0,n}$, thereby proving the proposition.

We will denote the intrinsic torus by T and always consider

$$\mathcal{M}_{0,n} \hookrightarrow T.$$

3.2 Selecting a Fan

We have already developed tropical compactification techniques for subvarieties of algebraic tori. Therefore, in order to compactify $\mathcal{M}_{0,n}$, we only choose a fan in the lattice $N_{\mathbb{R}}$. In what follows, we should keep in mind that the coordinates on $N_{\mathbb{R}}$ come from the coordinates of the natural dense torus in the the Plücker embedding, up an an overall $n-1$-dimensional additive translation.

We will describe the appropriate fan in two steps: first we specify an abstract complex obtained by gluing together cones along their faces, and second we embed this fan into the vector space $N_{\mathbb{R}}$. If we're aiming for the tropical compactification to be the traditional moduli space $\overline{\mathcal{M}}_{0,n}$, then we're expecting that the partially ordered set of faces in this proposed fan is "naturally" the partially ordered set of strata of $\overline{\mathcal{M}}_{0,n}$. Let us recall the basic language needed for this.

An *n-marked tree* is a finite tree G together with a marking function

$$\{1, \ldots, n\} \to V(G),$$

to the vertex set of G. A vertex of G is *stable* if the number of edges incident to that vertex, plus the number of markings at that vertex, is at least 3. An n-marked tree is stable if all vertices are stable.

The set of all stable n-marked trees forms a partially ordered set—if we contract an edge, or a series of edges, the resulting tree comes with a natural marking. These edge contractions define a natural partially ordered set structure on the set of all such trees. We denote it $\mathsf{A}_{0,n}$. It is, in a natural way, the same partially ordered set structure as the one on the set of generic points of strata of $\overline{\mathcal{M}}_{0,n}$.

We will realize the partially ordered set above as the set of faces of a fan. A *n-marked stable rational tropical curve* is a *stable n-marked tree* together with a *length* function:

$$\ell(-) : E(G) \to \mathbb{R}_{>0}.$$

The set of all such tropical curves whose underlying marked tree is a fixed stable n-marked tree is therefore naturally parameterized by $\mathbb{R}_{>0}^{E(G)}$. As we move to the faces of this cell where a subset of the lengths go to 0, the limiting face can naturally be viewed as a cell corresponding to a contraction of G, at exactly the set of edges whose lengths are set to 0.

The set of n-marked stable rational tropical curves, with this cone complex, will be the cone complex of our fan:

$$\Sigma_{0,n} = \{n\text{-marked stable rational tropical curves}\}.$$

Now we embed $\Sigma_{0,n}$ into the vector space $N_{\mathbb{R}}$. Associated to a metric tree $(\Gamma, p_1, \ldots, p_n)$ is a tuple of $\binom{n}{2}$ distances that allows us to embed

$$\Sigma_{0,n} \hookrightarrow \mathbb{R}^{\binom{n}{2}}.$$

It can be checked that projecting from $\mathbb{R}^{\binom{n}{2}}$ onto $N_{\mathbb{R}}$ gives us a fan that we want, where the projection is by quotienting along the linear subspace

$$\mathbb{R}^n \to \mathbb{R}^{\binom{n}{2}}; \quad \underline{a} \mapsto (a_i + a_j)_{ij}.$$

We obtain a toric variety $X_{0,n}$ and a sequence of maps

$$\mathcal{M}_{0,n} \hookrightarrow T \hookrightarrow X_{0,n}.$$

Putting all the pieces together, we are led to this beautiful theorem.

Theorem 3.2.1 *The closure of $\mathcal{M}_{0,n}$ is a tropical compactification. It is proper, smooth with simple normal crossings boundary, and isomorphic to space $\overline{\mathcal{M}}_{0,n}$. It meets all torus orbits of $X_{0,n}$ and does so in the expected dimension. In particular, the tropicalization of $\mathcal{M}_{0,n}$ is $\Sigma_{0,n}$.*

Let us focus more for a moment on the fact that $\overline{\mathcal{M}}_{0,n}$ is a tropical compactification. Recall T is the dense torus of $X_{0,n}$. We examined the morphism

$$\overline{\mathcal{M}}_{0,n} \to [X_{0,n}/T].$$

Without focusing on the precise stack structure on the latter, we note that as a topological space it merely consists of one point for each cone of $\Sigma_{0,n}$; it is a partially ordered set, namely the set of faces of $\Sigma_{0,n}$. The preimage of that point is a locally closed subspace of $\overline{\mathcal{M}}_{0,n}$ of codimension equal to the dimension of this cone.

I will not give a proof of the theorem above, but I will explain a route to a complete proof.

Remarks on the proof

(i) We first address the properness. In order to guarantee properness, we need to show that the tropicalization map

$$\mathcal{M}_{0,n}(K) \to \mathbb{R}^{\binom{n}{2}-n}$$

is contained in $\Sigma_{0,n}$. In fact, they are equal. To see this, one needs equations for $\mathcal{M}_{0,n}$, but these are hard to come by. An easier problem is to obtain equations for $G(2,n)^\circ$ in its Plücker embedding. Equations can be found in any standard textbooks. The fact that $G(2,n)^\circ$ is acted upon by a torus of dimension $n-1$ implies immediately that its tropicalization is acted upon by a vector space $\mathbb{R}^{n-1}$. One obtains the requisite statement by analyzing the equations and passing to the quotient. The result is addressed, beautifully and in detail, in the book of Maclagan–Sturmfels [MS15, Chapter 6].

(ii) We next note that the theorem on existence of tropical compactifications does not guarantee that a single fan exists such that the closure of $\mathcal{M}_{0,n}$ is simple normal crossings. Indeed, that claim implies that

$$\overline{\mathcal{M}}_{0,n} \times T \to X_{0,n}$$

is *smooth* rather than merely flat/equidimensional. Some elementary definition chasing shows that the condition of being SNC is equivalent to first, $X_{0,n}$ being smooth, which is combinatorial and clear and (ii) the *initial degenerations* of $\mathcal{M}_{0,n}$ being smooth. The initial degenerations are each subvarieties of tori obtained by manipulating the equations for $\mathcal{M}_{0,n}$. The input for such a manipulation is a point w in $\Sigma_{0,n}$. Fixing w, each equation for $\mathcal{M}_{0,n}$ is replaced by its initial terms i.e. lowest weight terms for weight determined by w. In this way, we obtain a scheme for each w. One can then immediately calculate from the defining equations for the Plücker embedding that we have the requisite smoothness.

(iii) Finally, to show the isomorphism with the standard moduli space $\overline{\mathcal{M}}_{0,n}$ one needs to produce a morphism. A simple way to do this to use the universal property and Zariski's main theorem. But amusingly we will never need this!

□

The space $\Sigma_{0,n}$ knows almost everything there is to know about $\overline{\mathcal{M}}_{0,n}$.

Theorem 3.2.2 *Consider the inclution* $\iota : \overline{\mathcal{M}}_{0,n} \hookrightarrow X_{0,n}$. *The pullback morphism*

$$\iota^\star : \mathsf{CH}^\star(X_{0,n}) \to \mathsf{CH}^\star(\overline{\mathcal{M}}_{0,n})$$

is an isomorphism of rings.

Stable Maps and Logarithmic Stable Maps

4

We summarize the discussion of tropical compactification for $\mathcal{M}_{0,n}$. We have five objects in play.

1. The open moduli space $\mathcal{M}_{0,n}$, embedded in its intrinsic torus T.
2. The tropical moduli space $\Sigma_{0,n}$, embedded in the cocharacter vector space $N_{\mathbb{R}}$ of T.
3. The toric variety $X_{0,n}$ with dense torus T.
4. The tropical compactification of $\mathcal{M}_{0,n}$ in $X_{0,n}$ which is identified with $\overline{\mathcal{M}}_{0,n}$.
5. A partially ordered set $A_{0,n}$—simultaneously of faces of $\Sigma_{0,n}$ under inclusion, of torus orbits of $X_{0,n}$ under orbit-closure containment, of strata $\overline{\mathcal{M}}_{0,n}$ under strata-closure containment, of topological types of n pointed stable genus 0 curves, and of combinatorial type of n-pointed genus 0 abstract tropical curves.

4.1 Curves in the Algebraic Torus

We recall that our goal was to calculate the degree of the map

$$\mathrm{ev} : \mathcal{M}_n(\mathbb{P}^2, d) \to (\mathbb{P}^2)^n,$$

sending a morphism from a marked rational curve to $\mathbb{P}^2$, to the tuple of images of the markings.

In order to achieve this, we first tweak the problem to put it in the striking zone of tropical compactification techniques. Fix coordinates on $\mathbb{P}^2$ with coordinate hyperplanes

H_0, H_1, H_2. We will consider maps, called *non-degenerate logarithmic maps* with transverse tangency conditions of degree d to $\mathbb{P}^2$,

$$f : (\mathbb{P}^1, p_1, \ldots, p_d, q_1, \ldots, q_d, r_1, \ldots, r_d, s_1, \ldots, s_n) \to (\mathbb{P}^2, H_0, H_1, H_2),$$

of pairs, with the property that

$$f^{-1}(H_0) = \sum p_i, \quad f^{-1}(H_1) = \sum q_i, \quad f^{-1}(H_2) = \sum r_i.$$

Two such maps are equivalent if they differ by reparameterizing the source; equivalently, we can take the first three points to be 0, 1, and ∞ without any equivalence relation. There is a moduli space $\mathcal{M}_{0,n}^{\log}(\mathbb{P}^2, d)$ of such nondegenerate logarithmic maps. The following lemma is straightforward.

Lemma 4.1.1 *The natural morphism from the space $\mathcal{M}_{0,n}^{\log}(\mathbb{P}^2, d)$ to $\mathcal{M}_n(\mathbb{P}^2, d)$ remembering the underlying map and the n marked points $s_1, \ldots, s_n$ is dominant and has degree $(d!)^3$.*

Proof The locus in $\mathcal{M}_n(\mathbb{P}^2, d)$ where the curve meets H_0, H_1, H_2 transversely at a set of reduced points away from the coordinate points is open and dense. Labeling the points gives the degree $(d!)^3$ cover. □

Since all the preimages of H_i are marked, evaluation at the points $s_1, \ldots, s_n$ lies in $\mathbb{G}_m^{2n}$, i.e. n copies of the dense torus of $\mathbb{P}^2$:

$$\mathrm{ev} : \mathcal{M}_{0,n}^{\log}(\mathbb{P}^2, d) \to \mathbb{G}_m^{2n}.$$

We are now interested in computing the degree of this map.

4.2   Curves in Toric Varieties

While our discussion above has been about maps to $\mathbb{P}^2$, it is no more difficult to solve a more general problem. Let X be a smooth and projective toric variety and let D be the toric boundary, with divisorial components $D_1, \ldots, D_m$. We fix an integer matrix (c_{ij}) of size $s \times m$, where s will be the number of marked points in what follows. We will study maps

$$f : (\mathbb{P}^1, q_1, \ldots, q_s) \to (X, D),$$

with the properties that (i) the preimage $f^{-1}(D)$ is precisely the set of points $\{q_i\}$, and (ii) the order of vanishing of the section cutting out D_j along the point q_i is c_{ij}. The moduli space $\mathcal{M}(X|D)$ of such maps is non-compact, and we will see below that it is closely related to the space $\mathcal{M}_{0,n}$. Note that we have suppressed the integers c_{ij} from the notation. We are interested in the degrees of the evaluation maps

$$\text{ev} : \mathcal{M}^{\log}(X|D) \to \mathbb{G}_m^N.$$

The calculation of the degree of ev above is of course a special case.

4.3 One Parameter Degenerations and Tropical Curves

We compactify this space by a similar strategy to the $\overline{\mathcal{M}}_{0,n}$ case above—by finding an appropriate set of tropical objects, assembling them into a cone complex/fan, and taking closure in an appropriate toric variety.

The way we will produce tropical data is as follows. Given a map f : $(\mathbb{P}^1, q_1, \ldots, q_s) \to (X, D)$ as above, we can puncture out the divisor D and the points q_i on C to get

$$C^\circ \to T$$

where C is a punctured smooth genus 0 curve. If the data are defined over the Puiseux series field K from before we extract from this a *tropical map*:

$$\Gamma \to N_{\mathbb{R}},$$

where Γ is a stable n-marked tropical curve. In order to get Γ we use what we already know about the moduli space $\overline{\mathcal{M}}_{0,n}$. First, mark the punctured points $C \setminus C^\circ$ and consider the limiting family in $\overline{\mathcal{M}}_{0,n}$. The family gives rise to a metric graph Γ: we take the dual graph of the limiting curve and decorate each edge with the rate at which the edge smooths. Precisely, for each edge e, there is a node q_e in the special fiber, which locally looks like

$$xy = t^\ell$$

essentially by definition of a nodal curve. The edge length is taken to be ℓ. A good exercise is to check that this is well-defined. In order to construct the map, we use the same definition as before of coordinatewise valuations.

In the compactification that we will construct below, the space $\mathcal{M}(X|D)$ will form the interior, and the geometrically simplest part. It parameterizes

$$(C, q_1, \ldots, q_s) \to (X, D)$$

as above. The moduli space of logarithmic stable maps is a compactification of this space—a typical point in it will determine a map from a nodal curve to X as above, *together* with a tropical map as above.

4.4   Constructing the Space

Let us now construct the space. Our starting point is the space $\mathcal{M}(X|D)$ outlined above, with the contact orders (c_{ij}) fixed. Suppose T' is the dense torus of X. It is very good to keep the following example in mind. Take $X = \mathbb{P}^1$ with toric boundary D being 0 and ∞. In this case, fixing the contact orders is simply attaching an integer number a_i associated to p_i. Note that the sum of all a_i must be zero for such a map to exist. It's very easy to write down a map:

$$z \mapsto \lambda \cdot \prod_{i=1}^{n}(z - p_i)^{a_i}, \quad \lambda \in \mathbb{C}^*.$$

In fact, this is true in general, and is a basic exercise.

Proposition 4.4.1 *Let (X, D) be a pair consisting of a toric variety and its toric boundary divisor. Consider the space $\mathcal{M}(X|D)$ of maps*

$$(\mathbb{P}^1, p_1, \ldots, p_n) \to (X, D)$$

with fixed contact orders c_{ij} of each marked point p_i with each divisor D_j. The forgetful map

$$\mathcal{M}(X|D) \to \mathcal{M}_{0,n}$$

is a torsor under the action of the group $T' = X \smallsetminus D$.

A simple calculation using the excision exact sequence shows that the Picard group of $\mathcal{M}_{0,n}$ is trivial, and so there are no nontrivial torus torsors. In particular, the moduli space of such maps is therefore abstractly isomorphic to

$$\mathcal{M}(X|D) = \mathcal{M}_{0,n} \times T'.$$

The theory of logarithmic stable maps will compactify this space. As before, we embed

$$\mathcal{M}(X|D) \hookrightarrow T \times T'$$

where T is the intrinsic torus of $\mathcal{M}_{0,n}$ and T' is the target torus. In order to compactify, we simply need to put a fan structure on this target. In fact, tropicalization respects products so we already know the support of this fan is

$$\mathrm{trop}(\mathcal{M}(X|D)) = \Sigma_{0,n} \times N'_{\mathbb{R}}.$$

In analogy to what was done for $\overline{\mathcal{M}}_{0,n}$, we will place a fan structure that mimics the stratification that we want from our toric varieties.

A tropical stable map to $N'_{\mathbb{R}}$ is an abstract tropical curve Γ together with a map

$$\Gamma \to N'_{\mathbb{R}}$$

that is piecewise linear and satisfies the balancing condition. It is a nice exericse to understand the structure of the set of all such tropical maps. We define the contact orders of such a map to be the tuple of edge directions of the ends of Γ, together with their slope.

Proposition 4.4.2 *Let Γ be an abstract tropical curve. Consider the set Δ of tropical stable maps*

$$\Gamma \to N'_{\mathbb{R}}$$

with fixed asymptotic slopes along the marked ends. The map

$$\Delta \to \Sigma_{0,n}$$

obtained by forgetting the map and remembering the underlying stable rational tropical curve is a torsor under the additive group $N'_{\mathbb{R}}$.

4.5 Tropical Compactification

The moduli space $\mathcal{M}_{0,n} \times T'$ has been embedded in a torus. Over this moduli space, there is a universal family $\mathcal{C}$ whose fiber is the *punctured curve* on the $\mathcal{M}_{0,n}$-factor. The latter can be identified with $\mathcal{M}_{0,n+1} \times T'$. By construction, there is a universal map from this construction to T'. All three spaces are therefore embedded in algebraic tori.

In order to compactify the moduli space $\mathcal{M}(X|D)$ and its universal curve, we will now select fan structures for these ambient tori. By the theory of tropical compactifications, we need to select fan structures on

$$\Gamma^{\mathrm{univ}} = \Sigma_{0,n+1} \times N'_{\mathbb{R}},$$

for $\Sigma_{0,n} \times N'_{\mathbb{R}}$, and for $N'_{\mathbb{R}}$. The choice for $N'_{\mathbb{R}}$ is clear from the choice of X, so will be taken to be Σ_X.

For the rest, we examine: what do we want from a moduli space of logarithmic stable maps? At the tropical level, we have a diagram:

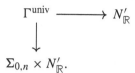

It is easy to see that these spaces live inside the natural tori that contain $\mathcal{M}_{0,n} \times T'$, its universal curve, and (trivially) the dense torus T' of X. We want to compactify the corresponding algebraic diagram

$$
\begin{array}{ccc}
\mathcal{C}^\circ & \longrightarrow & T' \\
\downarrow & & \\
\mathcal{M}_{0,n} \times T. &
\end{array}
$$

By arguments we used for $\mathcal{M}_{0,n}$, in order to compactify, we play the game of putting fan structures on the tropical diagram and take closures. The fan structure Σ_X on $N'_{\mathbb{R}}$ is given to us. In order for us to compactify, reasonably we want two further conditions:

1. The compactified moduli spaces should have a map to X;
2. The universal curve should be flat.

In order to achieve the first, we take the common refinement of Σ and the universal curve structure. This is actually easy—it is the fiber product in the category of fans! Next, we obtain a refinement $\widetilde{\Gamma}^{\text{univ}}$ of the fan of the universal curve mapping to $\Sigma_{0,n} \times \Sigma_X$. The flatness is visible in a simple way—every cone of the source should surject onto a cone of the target. We arrange for the smallest modification that makes this possible—noting that there is one by the exercise above—and we're there:

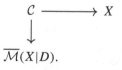

It can be shown that the resulting space coincides with the space of logarithmic stable maps, constructed in a much more general context, by Abramovich–Chen [AC14, Che14] and Gross–Siebert [GS13].

4.6   **Correspondence Theorems**

We now prove the simplest form of tropical correspondence theorem. Given evaluation classes:

$$\mathrm{ev} : \mathcal{M}(X|D) \to X^n.$$

In fact, by the universal property of the intrinsic torus, this factors via a toric variety

$$\overline{\mathcal{M}}(X|D) \hookrightarrow Y_{\mathrm{big}} \to X^n.$$

We are interested in enumerative invariants given, for example, by the degree of this composite map. Let us focus on this particular example, which covers the case of the numbers N_d that we set out to understand. The case of more general "incidence constraints", namely numbers obtained by pulling back general cohomology classes from X^n, can be treated similarly.

Goal *Assume that the dimension of $\overline{\mathcal{M}}(X|D)$ is equal to the dimension of X^n. Calculate the degree of the morphism* $\mathrm{ev} : \overline{\mathcal{M}}(X|D) \to X^n$.

The advantage of compactifying the space, as we have, is that we can approach the calculation of the degree from an intersection theory point of view. Let us choose torus invariant divisors $H_1, \ldots, H_e$ in X^n that intersect transversely at a point. There are choices in doing this: the fan of X^n is Σ_X^n, and if we choose any smooth, maximal dimensional cone, the divisors corresponding to the rays of that cone give the requisite $H_1, \ldots, H_e$.

We are now asking to calculate the degree of the 0-dimensional homology class:

$$\left(\mathrm{ev}^\star(H_1) \cup \cdots \cup \mathrm{ev}^\star(H_e) \right) \cap [\overline{\mathcal{M}}(X|D)].$$

Note that the expression here makes sense: the cup product is defined regardless of the singularities of $\overline{\mathcal{M}}(X|D)$, and the cap product only requires the moduli space to be irreducible.

Let us now contemplate how to actually calculate one of these divisors $\mathrm{ev}^\star(H_i)$. The divisor H_i is a torus invariant Cartier divisor on the space X^n and therefore corresponds to a piecewise linear function on Σ_X^n. Precisely, it is the function whose slope along the ray associated to H_i is 1 and whose slope along all other rays is 0. In order to perform the pullback, we first pullback along

$$Y_{\mathrm{big}} \to X^n.$$

This morphism is equivariant, and the map on fans has a beautiful description. Recall that the toric fan of Y_{big} was chosen already to be the moduli space of tropical curves equipped with a map to Σ_X. These tropical curves have "marked ends" corresponding to the marked points at which we are evaluating. The map on fans corresponding to $Y_{\text{big}} \to X^n$ just evaluates the position at each of these marked ends tropically.

We may now pullback H_i along $Y_{\text{big}} \to X^n$ by composing the associated piecewise linear function. In principle, one can now conclude as follows: write each pullback of H_i as a linear combination of boundary divisors, and then interpret this expression in terms of the boundary divisors of $\overline{\mathcal{M}}(X|D)$.

We can do even better. Let us suppose that, by some miracle, the map $Y_{\text{big}} \to X^n$ were flat. There is a purely combinatorial criterion for this—every cone in the fan Σ_{big} of Y_{big} should map surjectively onto a cone of Σ_X^n. If this holds, then in fact there is a completely combinatorial way to calculate the degree. It is achieved in the following steps:

1. Fix a cone σ of Σ_X^n in the target.
2. Enumerate the cones of the fan Σ_{big} that map surjectively onto σ.
3. For each cone $\tilde{\sigma}_j$ that maps to σ, calculate the absolute value m_j of the determinant of the associated linear map.
4. Add up the multiplicities m_j over all such cones.

Assuming the flatness condition, it is a simple exercise in using the toric dictionary to prove this result. The flatness condition can occur, but even if it doesn't, we can appeal to the *strong* semistable reduction theorem, due to Adiprasito et al. [ALT18], which says that $Y_{\text{big}} \to X^n$ can be replaced by maps of blowups $Y_{\text{big}}^{\dagger} \to X^{n,\dagger}$ that are smooth, and where the map is, in addition, flat.

We have therefore achieved our goal of a purely combinatorial formula for the degree of ev. By using this, and the finer analysis of the tropical side in Part III, we are led to the proof of the correspondence theorem of Mikhalkin and Nishinou–Siebert [Mik05, NS06], see also Theorem 9.5.4 in Chap. 9, Sect. 9.5.

Cheat Codes for Logarithmic GW Theory

5

The goal of this final section is to give the reader a set of cheat codes to navigate logarithmic GW theory. The original paper by Gross–Siebert is still essentially the modern presentation of these ideas, and one cannot avoid grappling with it forever [GS13]. Still, by now illustrating the key general features of the spaces we have constructed in our nice context—genus 0 curves in a toric variety—will give the reader a guide to this subject.

5.1 The Open Moduli Space

The target in logarithmic Gromov–Witten theory is a pair (X, D) consisting of a smooth projective variety X together with a simple normal crossings divisor D. Typical examples include the case where X is a smooth toric variety and D is a subset of its toric boundary, and the case where X is $\mathbb{P}^n$ and D is a generic union of hypersurfaces.

Let us fix a target (X, D) and let us assume D consists of irreducible components $D_1, \ldots, D_r$. For simplicity, we will assume that all intersections of subsets of these components are connected.

A *non-degenerate logarithmic map* is a smooth pointed curve $(C, p_1, \ldots, p_n)$ together with map:

$$F : (C, p_1, \ldots, p_n) \to (X, D)$$

of *pairs*—the preimage of D is a union of some (but not necessarily all) of the points p_i. Given a point p_i and a divisor component D_j, there is a well-defined *tangency order*—pull back the defining equation for D_j to p_i and calculate the order of vanishing.

The discussion so far sets up the numerical data of the problem: we fix the genus g of the curves C, the number n of marked points, the class $F_*[C]$ in $H_2(X; \mathbb{Z})$, and the

R. Cavalieri et al., *Tropical and Logarithmic Methods in Enumerative Geometry*, Oberwolfach Seminars 52, https://doi.org/10.1007/978-3-031-39401-0_5

contact order matrix A, whose entry a_{ij} is the tangency order above. Let Γ package the full set of numerical data. The moduli space $\mathcal{M}_\Gamma(X|D)$ is a Deligne–Mumford stack with quasi-projective coarse space.

5.2   Logarithmic Structures: The Bare Minimum

Suppose we are given a pair (Y, E) of a smooth projective variety and normal crossings divisor E. The sheaf of regular functions $\mathcal{O}_Y$ on Y contains certain distinguished elements. Indeed, each irreducible component E_i of E gives rise to a line bundle $\mathcal{L}_i$ and a section s_i, well-defined up to scalar, that cuts out E_i. If we look locally enough, this s_i is just an element of the structure sheaf.

Thus, loosely speaking, a pair (Y, E) has not only a notion of "polynomial function", i.e. elements of $\mathcal{O}_Y$ but also a notion of *monomial functions*. Note that monomials can be multiplied, but typically cannot be added, to get more monomials. In other words, the natural structure on monomials is that of a *monoid* or *semigroup*.

In logarithmic geometry, *all* objects come with a notion of monomial function, and the category of logarithmic schemes has many of the familiar properties of traditional schemes, including fiber products. The first piece of intuition that breaks, however, is that *one can have more monomials than polynomials*. Let us now give the precise definition.

Definition 5.2.1 Let Y be a scheme. A logarithmic structure on Y is a sheaf of monoids M_Y together with a map of monoids (where $\mathcal{O}_Y$ is a multiplicative monoid)

$$\alpha : M_Y \to \mathcal{O}_Y,$$

such that $\alpha^{-1}(\mathcal{O}_Y^*)$ maps isomorphically under α to $\mathcal{O}_Y^*$. A logarithmic scheme is a scheme together with a logarithmic structure. It will be denoted (Y, M_Y).

Let us unpack this definition with a series of examples.

Example 5.2.2 The best example to keep in mind is a simple normal crossings pair (Y, E). On an open set U the sheaf M_Y evaluates to the subsheaf of $\mathcal{O}_Y(U \smallsetminus E)$ consisting of invertible functions. In other words, the monomials are locally monomials in the defining equations of the components of E.

Example 5.2.3 The next best example to keep in mind is an affine toric variety U_P associated to a toric monoid P, with coordinate ring $\mathbb{C}[P]$. In this case, given an element $m \in P$, there is an associated character χ^m, obtained by viewing $P \subset M$, where M is the character lattice. Thus we already see that there is a reasonable notion of monomial function on a toric variety (provided we remember the *way* in which U_P is a toric variety). In fact, there is a canonical logarithmic structure on U_P which we canote M_P. If we evaluate this sheaf on U_P itself, the resulting monoid is $\mathbb{C}^\star \oplus P$, where an element (λ, m) is mapped (by α above) to $\lambda \chi^m$.

Example 5.2.4 The richest example to keep in mind is $\overline{\mathcal{M}}_{g,n}$—the moduli space of stable genus g curves with n marked points. The logarithmic structure here is defined on the étale topology, by asserting that the divisor of singular curves, which has simple normal crossings on an étale cover, gives the monomials.

Let us explain this in practical terms. We are looking at a small open neighborhood around a moduli point $[C, p_1, \ldots, p_n]$. By passing to an étale cover U, we can label the nodes of C as $q_1, \ldots, q_r$. At a generic point in this neighborhood U the corresponding curve will be smooth. However, there are divisors in U where the node q_i persists. The functions cutting out these divisors give rise to monomial functions in U. These endow $\overline{\mathcal{M}}_{g,n}$ with a logarithmic structure.

Finally, let us look at one of the stranger, but central, examples in the subject.

Example 5.2.5 Let Y be the scheme $\mathrm{Spec}(\mathbb{C})$ and choose any monoid P with a neutral element 0 and no other invertible elements. For the purposes of concreteness, the reader can keep in mind $\mathbb{N}$ or $\mathbb{N}^r$. A sheaf of monoids on X is just the data of a monoid, and the sheaf $\mathcal{O}_Y^\star$ is the sheaf $\mathbb{C}^\star$. We know that the logarithmic structure M_Y needs to contain $\mathbb{C}^\star$, so the simplest way to involve P is to declare $M_Y = \mathbb{C}^\star \oplus P$. The map

$$\alpha : M_Y \to \mathcal{O}_Y$$

sends (λ, u) to λ if u is 0 and to 0 otherwise. Although this is strange, it has a nice interpretation. Let us set $P = \mathbb{N}$. Then we should think of the logarithmic scheme (Y, M_Y) as a neighborhood of the origin in the toric variety $\mathbb{A}^1$, where $\mathbb{A}^1$ is treated as a logarithmic scheme as in the previous examples. Then we interpret (λ, n) as being the element λt^n, and "set $t = 0$", but in the sheaf M_Y we remember the order vanishing n of this function. Of course, this function has to become 0 in the structure sheaf $\mathcal{O}_Y$, and here we see that M_Y is *not a submonoid* of $\mathcal{O}_Y$. in other words, *there are more monomials than polynomials!* The monomials here are the elements of $\mathbb{N}$, which we interpret as "t^n", even though as functions on $\mathrm{Spec}(\mathbb{C})$ they are all 0.

In the world of toric varieties, once a dense torus is chosen, there is not only a notion of monomial function, but also a notion of "pure monomial" function, namely those monomials with coefficient 1. The set of pure monomials form a *submonoid* of all monomial functions. However, this is an accident that comes from the group structure on the torus. In more general contexts, one cannot pin down the notion of a pure monomial globally on a scheme. The right structure on "pure monomials", is not that of a submonoid but a quotient monoid. This notion generalizes well to the logarithmic setting as we now explain.

Given a logarithmic scheme (Y, M_Y) there is a fundamental short exact sequence:

$$0 \to \mathcal{O}_Y^\star \to M_Y^{\mathrm{gp}} \to \overline{M}_Y^{\mathrm{gp}} \to 0.$$

The term on the far right is called the *characteristic abelian sheaf* of the logarithmic structure. The image of M_Y is called the characteristic monoid sheaf, or simply the characteristic sheaf.

Example 5.2.6 Let (Y, E) be a simple normal crossings pair and let M_Y be the logarithmic structure discussed above. Then the characteristic sheaf $\overline{M}_Y$ is a constructible sheaf. At a point $y \in Y$, it is isomorphic to $\mathbb{N}^{e(y)}$, where $e(y)$ is the number of divisor components of E passing through y. In other words, its rank measures "how deep into the boundary" the point y lives.

A *morphism* of logarithmic schemes $(Z, M_Z) \to (Y, M_Y)$ is a morphism $f : Z \to Y$ of schemes, a morphism of sheaves of monoids $f^\flat : f^\star M_Y \to M_Z$ that is compatible in the natural way. To spell this out, α tells us how to take a monoid element and think about it as a function. As a result, such elements can first be turned into functions and then pulled back, or first be pulled back, and then turned into functions. The result should be the same.

Three simple examples give a basic illustration of morphisms.

1. If (Z, F) and (Y, E) are simple normal crossings pairs, as above, a morphism $Z \to Y$ is logarithmic precisely when the preimage of E is contained in F.
2. Every equivariant morphism between toric varieties is a logarithmic morphism, with the canonical toric logarithmic structures. If X is a toric variety, then the translation map $X \to X$ induced by multiplication by a fixed element of the dense torus is *also* logarithmic. Together, under composition, these give rise to *all* logarithmic morphisms between toric varieties—with their canonical logarithmic structures.
3. Building off the first example, the inclusion of a toric boundary divisor into an ambient toric variety is *not* a logarithmic morphism.

The reader is encouraged to make themselves comfortable with those three examples. But, from this stage on, the construction of the category of logarithmic schemes is a formal exercise, essentially identical to that of schemes. We make two crucial final comments though:

1. We typically work with on *fine and saturated* logarithmic schemes, which is a niceness condition on the stalks $\overline{M}_{Y,y}$ of the characteristic sheaf. We demand that all these monoids define *normal toric varieties*. The resulting objects form a full subcategory of logarithmic schemes.
2. The category of all logarithmic schemes does admit fiber products, and the full subcategory of fine and saturated logarithmic schemes also admits fiber products.

However, they do not coincide, and this is one of the most important and subtle points of the whole theory! Given two fine and saturated logarithmic schemes Z and Y over a logarithmic base B, the "all logarithmic" fiber product is a logarithmic structure on the fiber product of the underlying scheme: $Y \times_B Z$. But the fine and saturated fiber product typically has a finite, non-surjective morphism to this $Y \times_B Z$.

5.3   The Compact Moduli Space

The main construction in the papers [AC14, Che14, GS13] is actually very simple to explain. We will record it, and then explain how one might think about it.

The main definition is that of a family of logarithmically smooth curves. One can given an intrinsic definition, but we will cheat a little.

The stack $\mathfrak{M}_{g,n}$ parameterizes families of genus g nodal curves with n marked points *without any stability condition*. It is an Artin stack, rather than a Deligne–Mumford stack, and it is not separated. While a family of smooth curves over a punctured disk always has a stable limit, one can attach arbitrary trees of rational curves to that stable limit; the stack $\mathfrak{M}_{g,n}$ does not attempt to distinguish between all these limits. As a result, psychologically, it is a large object. Nevertheless, it shares many of the properties of $\overline{\mathcal{M}}_{g,n}$. Precisely, it is smooth of dimension $3g - 3 + n$ and contains a divisor parameterizing singular curves. In the appropriate topology, namely the topology of smooth (as opposed to just étale) maps, $\mathfrak{M}_{g,n}$ is a normal crossings pair and therefore has a logarithmic structure.

Similarly, its universal curve $\mathfrak{C} \to \mathfrak{M}_{g,n}$ also has a logarithmic structure, and the morphism is logarithmic.

Definition 5.3.1 Let (S, M_S) be a logarithmic scheme. A logarithmic curve over (S, M_S) is a logarithmic morphism $S \to \mathfrak{M}_{g,n}$.

Given this morphism, we pull back the universal logarithmic curve to obtain a family of actual curves over S, with logarithmic structures, etc. We will denote such a family C/S. Starting now, we will start suppressing the sheaf M_X from the notation.

Definition 5.3.2 Let (X, M_X) be a logarithmic scheme. A *prestable logarithmic map to* (X, M_X), over (S, M_S) is a logarithmic curve C/S together with a logarithmic map from the associated total space of curves to (X, M_X). Such a map is *stable* if the underlying scheme theoretic map $C \to X$ is stable.

The main theorem, proved by Abramovich–Chen and Gross–Siebert, is that there is an actual algebraic stack with a logarithmic structure parameterizing these objects—i.e. such that a family of stable logarithmic maps over a logarithmic scheme S is precisely the data of a logarithmic map from S to this object.

The space of logarithmic stable maps has several important properties. Let us specialize to the case where (X, M_X) is a smooth projective variety X with logarithmic structure from a simple normal crossings divisor D, as above. This is the main case of interest. We denote the moduli stack in this case $\overline{\mathcal{M}}(X|D)$.

1. The number of marked points, the arithmetic genus of the source curve, and the curve class associated to the stable map are all constant in flat families, as in the traditional case.
2. Associated to each marked point p_i and divisor component D_j is a *contact order*. Indeed, after passing to local neighborhoods, we can assume that D_j is cut out by f_j which lies in the image of $M_X \to \mathcal{O}_X$. Choosing a lift e_j to M_X, we can pull back e_j to obtain an element of the stalk M_{C,p_i}. Its image in $\overline{M}_{C,p_i}$ is a natural number, which is called the contact order.
3. The contact order is a non-negative integer, it agrees with the tangency order for non-degenerate maps, and it is locally constant in flat families.
4. If Γ denotes the full set of numerical data as in the previous discussion, the space $\overline{\mathcal{M}}_\Gamma(X|D)$ is a proper Deligne–Mumford stack. It is potentially singular, non-reduced, reducible, etc.

5.4   The Space $\overline{\mathcal{M}}_\Gamma(X|D)$

When $(X|D)$ is a smooth and projective toric variety with logarithmic structure given by its toric boundary, and the genus g is equal to 0, we have constructed the space $\overline{\mathcal{M}}_\Gamma(X|D)$ already. We keep the notation because it is in fact an instance of the space of logarithmic stable maps.

Let us observe a couple of basic things about our space, that is easy to prove in the genus 0 case. Recall that we had, by construction, an embedding

$$\overline{\mathcal{M}}_\Gamma(X|D) \hookrightarrow Y_{\text{big}}$$

into an ambient toric variety.

Given a point $p \in \overline{\mathcal{M}}_\Gamma(X|D)$ we obtain two pieces of information: one purely geometric, and the other purely combinatorial:

1. The pure geometric data is a stable map from a pointed nodal curve $(C, p_1, \ldots, p_n) \to X$.
2. The pure combinatorics data is the cone dual to the stratum of Y_{big} in which p lies. It is an equivalence class of maps from tropical curves Γ to the fan Σ_X, where two maps are equivalent if they can be continuously deformed by adjusting edge lengths, without changing the cones in which the relative interiors of vertices or edges lie.

The data have to be *compatible*, which in particular implies the following. First, the dual graph of Γ has to coincide with the underlying graph of the tropical curves Γ. Second, if the generic point of a component C_v dual to a vertex v maps to a locally closed stratum dual to a cone $σ$, then the vertex v should map tropically to somewhere inside the cone $σ$.

While these are simple, they are unfortunately not the only things one has to keep in mind. There is more information hiding in the logarithmic structure, having to do with contact orders. But let us flag an important lesson:

> *A point of the space of logarithmic maps determines an ordinary stable map and a compatible equivalence class of tropical maps.*

In the next section, we explain the replacement in the general logarithmic case.

5.5 The Tropicalization

In toric geometry, one associates to each toric variety $(X|D)$ its fan $Σ_X$. The fan is typically viewed as a collection of cones, each embedded in the same vector space (of cocharacters). In logarithmic geometry, we think of the fan as a coarser object—it is just the abstract collection of cones with natural identifications between faces.

We want to build something similar for general fine and saturated logarithmic schemes. The basic idea is simple. On a toric variety $(X|D)$, the cones in the fan can be reconstructed from the set of coefficient-free, regular, monomial functions on torus invariant affines. Indeed, on each such affine, the functions for a monoid. The $\mathbb{R}_{\geq 0}$-dual of this monoid recovers the cone. We now do this on a general logarithmic scheme (X, M_X) by using the characteristic sheaf stalks $\overline{M}_{X,x}$.

We now do this honestly; the following construction is due to Gross–Siebert [GS13].

Construction 5.5.1 *Let (X, M_X) be a logarithmic scheme. For each point $x \in X$, we can consider the monoid $P_x : = \overline{M}_{X,x}$ and the dual cone $σ_x$. Consider open sets $U' \subset U$. Consider the restriction of an element f of $M_X(U)$ to $M_X(U')$. It may be that f does not lie in the kernel of the map to $\overline{M}_X(U)$, but does so on restriction to U'. As a result, we have natural specialization maps: if x is a specialization of y, there is a natural map of cones $σ_y \to σ_x$. The tropicalization of (X, M_X) is defined to be*

$$\text{trop}(X, M_X) = Σ_X = \varinjlim σ_x.$$

The construction certainly depends on the choice of the logarithmic structure, but this is often suppressed from the notation.

Example 5.5.2 Let (X, D) be a simple normal crossings pair with r components. Assume that all intersections of components of D are connected. Identify the subsets of $\{1, \ldots, r\}$

with the face of the cone $\mathbb{R}^r_{\geq 0}$ where precisely that subset of coordinates is nonzero. The tropicalization of the corresponding logarithmic scheme is the conical subcomplex of $\mathbb{R}^r_{\geq 0}$ corresponding to those subset of $\{1, \dots, r\}$ such that the corresponding intersection of $\overline{D}_i$ is nonempty.

As an exercise, you may want to show that the natural toric logarithmic structure on $\mathbb{P}^2$ and on $\mathbb{A}^3 \smallsetminus \{\underline{0}\}$ are the same—they are $\mathbb{R}^3_{\geq 0}$ minus the interior of the orthant.

The tropicalization is not quite functorial for maps of logarithmic schemes, but the issues surrounding this have to do with monodromy, and we will not discuss them here. The failure is mild, and can usually be dealt with in practice. Given a logarithmic morphism $X \to Y$, one expects there is a natural map $\Sigma_X \to \Sigma_Y$.

With these definitions, our lesson from the previous section continues to hold.

Let (X, D) be any simple normal crossings pair. Given a point of the moduli space $\overline{\mathcal{M}}_\gamma(X|D)$, this pointed determines an underlying stable map, together with an equivalence class of tropical curves and maps:

$$\Gamma \to \Sigma_X.$$

Unlike the toric case, we do not have such good control over the possibilities for such maps, but their structure is crucial. The tropical curves will now be dual graphs of higher genus curves, with decorations, but the general picture is the same. The stable map has topology that needs to be compatible with the combinatorial data.

The reader is encouraged to consult the details in [GS13] to unwind the details of this construction.

5.6   The Miraculous, Marvelous Artin Fans

The tropicalization is a very useful tool in logarithmic Gromov–Witten theory. However, it is useful to have the same data packaged in a more algebro-geometric fashion. This can be done by using a simple but ingenious idea of Abramovich–Wise [AW18].

Let Σ be any cone complex. For our purposes, it is plenty to think about $\mathbb{R}^r_{\geq 0}$. If (S, M_S) is a logarithmic scheme, we can consider the set-valued functor:

$$F_\Sigma : \mathbf{LogSch} \to \mathbf{Sets} \tag{5.1}$$

$$S \mapsto \mathrm{Hom}_{Cone}(\Sigma_S, \Sigma). \tag{5.2}$$

The functor is peculiar—its value on a logarithmic scheme depends only on the tropicalization of that scheme!

Like in ordinary moduli theory, one can ask if this functor is representable. It turns out, it is! The functor above is representable by a 0-dimensional Artin stack with logarithmic structure.

Definition 5.6.1 Let Σ be a cone complex. The representing functor F_Σ is called the *Artin fan* of Σ. If X is a logarithmic scheme, the Artin fan of X is the Artin fan associated to Σ_X.

The best examples to keep in mind are when the cone complex Σ is the fan of some toric variety.

Example 5.6.2 Let Σ be a fan associated to a toric variety X. The Artin fan of Σ is the stack quotient $[X/T]$.

Example 5.6.3 The Artin fan associated to $\mathbb{R}^r_{\geq 0}$ is $[\mathbb{A}^1/\mathbb{G}_m]^r$. We denote it $\mathcal{A}^r$.

Note that the same cone complex can be expressed as the fan of many different toric varieties, as we saw in the example previously. The quotient is nevertheless well-defined. Note that stacks of the form $[X/T]$ have a logarithmic structure in the smooth topology. Indeed, the map $X \to [X/T]$ is smooth, and therefore on smooth covers, we have a logarithmic structure.

5.7 Logarithmic Gromov–Witten Theory

By using Artin fans, one can give a very slick definition of a logarithmic map. Let C/S be a logarithmic curve. A logarithmic map to $\mathcal{A}^r$ is precisely the data of a diagram

$$
\begin{array}{ccc}
\Sigma_C & \longrightarrow & \mathbb{R}^r_{\geq 0} \\
\downarrow & & \\
\Sigma_S, & &
\end{array}
$$

where all arrows are maps of cone complexes. This defines an object of the stack of prestable maps $\mathfrak{M}_{g,n}(\mathcal{A}^r)$. Abramovich and Wise established the following beautiful result.

Theorem 5.7.1 *The stack $\mathfrak{M}_{g,n}(\mathcal{A}^r)$ is an Artin stack with logarithmic structure, having dimension $3g - 3 + n$.*

The result puts our earlier lessons on rigorous ground. A logarithmic map to $\mathcal{A}^r$ from a family of logarithmic curves is purely tropical information. Indeed, that's how we defined the Artin fan!

We can now say what the space of logarithmic stable maps is. Suppose that $(X|D)$ is a simple normal crossings pair. There is natural morphism

$$X \to \mathcal{A}^r.$$

By using this, we have:

Definition 5.7.2 The stack of logarithmic stable maps is the fiber product

$$\overline{\mathcal{M}}_{g,n}(X|D) := \overline{\mathcal{M}}_{g,n}(X) \times_{\mathfrak{M}_{g,n}} \mathfrak{M}_{g,n}(\mathcal{A}^r)$$

By imposing additional numerical conditions, such as contact orders and curve classes, one ends up with a finite type stack $\overline{\mathcal{M}}_\Gamma(X|D)$. A critical part of the theory is that there is a virtual fundamental class, and evaluation maps to X and its strata. These are enough to define and study logarithmic Gromov–Witten invariants [ACGS20a, ACGS20b, Che14, GS13]. There are other approaches to the subject, using the theory of expanded degenerations [Li01, Li02a, Ran22].

This is as far as we can get in these lecture notes. The purpose of this final section was to connect the work we did "honestly" for rational curves in toric varieties to the general theory. The author hopes that, together with the beautiful original papers in the subject and the detailed examples, this final section will give the reader an introduction to this beautiful theory.

Part II

Hurwitz Theory

Introduction

The goal of this part is to showcase the interactions of tropical and logarithmic geometry with Hurwitz theory, concerned with the enumeration of maps of Riemann surfaces.

The main question in Hurwitz theory dates all the way back to the late 1800s: how many maps of Riemann Surfaces does one have when fixing all the available discrete invariants? Over the last century, this question has experienced a wealth of translations (to topology, combinatorics, group theory, representation theory...) and found itself contributing to the most disparate areas of mathematics (integrable systems, mathematical physics, string theory...).

In this mini-course we focus on how Hurwitz theory interlaces with the geometry of moduli spaces of curves. The basic connection is that Hurwitz numbers are naturally interpreted as the degrees of appropriate branch morphisms among moduli spaces of covers and moduli spaces of target curves. After appropriately compactifying the moduli spaces, such degree is accessed through intersection theory.

The first manifestation of this phenomenon is the remarkable *ELSV formula*, that expresses simple Hurwitz numbers as Hodge integrals on the moduli spaces of curves. This formula was instrumental in explaining polynomial properties of simple Hurwitz numbers and in Okounkov-Pandharipande's proof of Witten's conjecture. A discussion of the ELSV formula, together with a brief sketch of how it may be proved through Atyiah-Bott localization is the punchline of the first lecture.

The second lecture is dedicated to how Hurwitz numbers may be accessed through *degeneration*: the number of covers of curves of fixed arithmetic genera can be computed by "shrinking a bunch of loops", and reducing to count maps among nodal curves of geometric genus zero, i.e. decomposing as the union of rational components. This technique leads to a combinatorial approach to Hurwitz theory which is well captured and organized by *tropical geometry*. In this lecture we explore several correspondence theorems between classical and tropical Hurwitz numbers, and we show how the tropical approach to double Hurwitz numbers makes their piecewise polynomiality and wall-crossing behavior very transparent. This part of the course connects with Part III, which in particular explores some of the foundations of tropical intersection theory. We conclude this lecture with an open conjecture, originally by Goulden-Jackson-Vakil, about finding an intersection theoretic formula similar to $ELSV$ for double Hurwitz numbers.

In the third lecture, we focus on how Hurwitz numbers (double and more) may be obtained as intersection numbers on moduli spaces closely related to the moduli space of curves. First we present an approach that uses the *double ramification cycle* to obtain an intersection theoretic formula for double Hurwitz numbers.

Next we turn our attention to a recent perspecive, which has been brough about by the development of logarithmic geometry: given a counting problem, logarithmic geometry gives access to two related moduli spaces, an algebro geometric one $\mathcal{M}$ and a tropical one $\mathcal{M}^{trop}$. One may in fact use the tropical one to define a birational modification $\overline{M}_{g,n}^+$ of the moduli space of curves in such a way that $\mathcal{M}$ intersects the boundary of $\overline{M}_{g,n}^+$ in a dimensionally transverse way. As explained in Part I, piecewise polynomial functions on the moduli space of tropical curves determine cohomology classes on $\overline{M}_{g,n}^+$. We show how we can construct some of these classes to compute Hurwitz numbers. This approach allows to extend the Hurwitz problem to moduli spaces of twisted differentials, circumventing the issue that these spaces lack a branch morphism.

These notes are meant to move fairly quickly through a lot of material, so they are by no means intending to be a complete reference. Many references are provided to help the interested reader. Our hope is to present a coherent and compelling story showcasing the development over many years in our understanding of Hurwitz theory.

Classical Hurwitz Theory and Moduli Spaces

6

In this first lecture we review some classical perspectives on Hurwitz numbers, and connect the problem of enumeration of maps of Riemann Surfaces with the geometry of tautological morphisms of appropriate moduli spaces. We discuss in particular the ELSV formula, which describes simple Hurwitz numbers as Hodge integrals, i.e. very special intersection numbers on moduli spaces of curves.

6.1   Hurwitz Numbers: Geometry

From a geometric point of view, Hurwitz numbers count the number of maps of Riemann surfaces with fixed discrete data and a fixed branch divisor.

Definition 6.1.1 (Geometry) Let $(Y, p_1, \ldots, p_r, q_1, \ldots, q_s)$ be an $(r + s)$-marked smooth Riemann Surface of genus h. Let $\underline{\eta} = (\eta_1, \ldots, \eta_s)$ be a vector of partitions of the integer d. We define the *Hurwitz number*:

© The Author(s), under exclusive license to Springer Nature Switzerland AG 2023
R. Cavalieri et al., *Tropical and Logarithmic Methods in Enumerative Geometry*,
Oberwolfach Seminars 52, https://doi.org/10.1007/978-3-031-39401-0_6

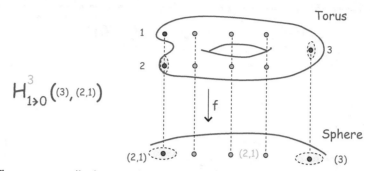

Fig. 6.1 The covers contributing to a given Hurwitz Number

$$H^r_{g \to h,d}(\underline{\eta}) := \textbf{weighted} \text{ number of} \left\{ \begin{array}{c} degree\ d\ covers \\ X \xrightarrow{\ f\ } Y\ such\ that : \\ \bullet\ X\ is\ connected\ of\ genus\ g; \\ \bullet\ f\ is\ unramified\ over \\ X \smallsetminus \{p_1, \dots, p_r, q_1, \dots, q_s\}; \\ \bullet\ f\ ramifies\ with\ profile\ \eta_i\ over\ q_i; \\ \bullet\ f\ has\ simple\ ramification\ over\ p_i; \\ \circ\ preimages\ of\ each\ q_i\ with\ same \\ ramification\ are\ distinguished\ by \\ appropriate\ markings. \end{array} \right\}$$

Each cover is weighted by the number of its automorphisms.

Figure 6.1 illustrates the features of this definition.

Remarks

(1) For a Hurwitz number to be nonzero, r, g, h and $\underline{\eta}$ must satisfy the Riemann Hurwitz formula

$$2g_X - 2 = d(2g_Y - 2) + \sum_{x \in X} (r_f(x) - 1).$$

The above notation is always redundant, and it is common practice to omit appropriate unnecessary invariants.

(2) The last condition $\circ$ was introduced in [GJV03] for the purpose of eliminating automorphism factors. These Hurwitz numbers differ by a factor of $\prod \text{Aut}(\eta_i)$ from the classically defined ones where such condition is omitted.

(3) One might want to drop the condition of X being connected, and count covers with disconnected domain. Such Hurwitz numbers are denoted by $H^\bullet$.

Example 6.1.2

-

$$H^0_{0\to 0,d}((d),(d)) = \frac{1}{d}$$

-

$$H^4_{1\to 0,2} = \frac{1}{2}$$

-

$$H^3_{1\to 0,2}((2),(1,1)) = 1$$

6.2   Hurwitz Numbers: Representation Theory

The problem of computing Hurwitz numbers is in fact a discrete problem and it can be approached using the representation theory of the symmetric group. A standard reference here is [FH91].

Given a branched cover $f : X \to Y$, a point y_0 not in the branch locus, and a labeling of the preimages $1, \ldots, d$, one can define a group homomorphism:

$$\varphi_f : \pi_1(Y \setminus B, y_0) \to \quad S_d$$
$$\gamma \quad \mapsto \sigma_\gamma : \{i \mapsto \tilde{\gamma}_i(1)\},$$

where $\tilde{\gamma}_i$ is the lift of γ starting at i ($\tilde{\gamma}_i(0) = i$). This homomorphism is called the **monodromy representation**, see Fig. 6.2.

Remarks

(1) A different choice of labelling of the preimages of y_0 corresponds to composing φ_f with an inner automorphism of S_d.
(2) If $\rho \in \pi_1(Y \setminus B, y_0)$ is a little loop winding once around a branch point with profile η, then σ_ρ is a permutation of cycle type η.

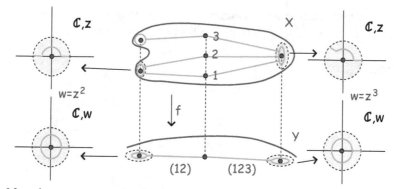

Fig. 6.2 Monodromy representation for the cover f

Viceversa, the monodromy representation contains enough information to recover the topological cover of $Y \smallsetminus B$, and therefore, by the Riemann existence theorem, the map of Riemann surfaces. To count covers we can count instead (equivalence classes of) monodromy representations. This leads to the second definition of Hurwitz numbers.

Definition 6.2.1 (Representation Theory) Let $(Y, p_1, \ldots, p_r, q_1, \ldots, q_s)$ be an $(r+s)$-marked smooth Riemann Surface of genus g, and $\underline{\eta} = (\eta_1, \ldots, \eta_s)$ a vector of partitions of the integer d:

$$H^r_{g \to h, d}(\underline{\eta}) := \frac{|\{\underline{\eta}\text{-monodromy representations } \varphi^{\underline{\eta}}\}|}{|S_d|} \prod \text{Aut}\eta_i, \qquad (6.1)$$

where an $\underline{\eta}$-**monodromy representation** is a group homomorphism

$$\varphi^{\underline{\eta}} : \pi_1(Y \smallsetminus B, y_0) \to S_d$$

such that:

- for ρ_{q_i} a little loop winding around q_i once, $\varphi^{\underline{\eta}}(\rho_{q_i})$ has cycle type η_i.
- for ρ_{p_i} a little loop winding around p_i once, $\varphi^{\underline{\eta}}(\rho_{p_i})$ is a transposition. $\star$ Im$(\varphi^{\underline{\eta}})$ acts transitively on the set $\{1, \ldots, d\}$.

Remarks

(1) To count disconnected Hurwitz numbers remove condition $\star$.
(2) Dividing by $d!$ accounts simultaneously for automorphisms of the covers and the possible relabellings of the preimages of y_0.
(3) $\prod \text{Aut}\eta_i$ corresponds to condition $\circ$ in Definition 6.1.1.

Note that the more natural problem from this perspective is the count of disconnected Hurwitz numbers, where the condition $\star$ is omitted. One may further translate the problem to a multiplication problem in the class algebra of the symmetric group, and exploit its semisimplicity to obtain closed formulas in terms of characters of the symmetric groups (Burnside formulas), see e.g. [CM16] for an elementary treatment. One may further consider all symmetric groups at the same time and express Hurwitz numbers as expectation values of certain operators on the Fock space, but this is another story, see [Joh14].

6.2.1 Disconnected to Connected: The Hurwitz Potential

The relationship between connected and disconnected Hurwitz numbers is systematized in the language of generating functions.

Definition 6.2.2 The **Hurwitz Potential** is a generating function for Hurwitz numbers. We present it with a redundand set of variables, keeping in mind that in almost all applications one makes a more efficient choice of the appropriate variables to mantain:

$$\mathcal{H}(p_{i,j}, u, z, q) := \sum H^r_{g \to 0, d}(\underline{\eta}) \, p_{1,\eta_1} \cdot \ldots \cdot p_{s,\eta_s} \frac{u^r}{r!} z^{1-g} q^d,$$

where:

- $p_{i,j}$, for i and j varying among non-negative integers, index ramification profiles. The first index i keeps track of the branch point, the second of the profile. For a partition η the notation $p_{i,\eta}$ means $\prod_j p_{i,\eta_j}$.
- u is a variable for unmarked simple ramification. Division by $r!$ reflects the fact that these points are not marked.
- z indexes the genus of the cover (more precisely it indexes the euler characteristic, which is additive under disjoint unions).
- q keeps track of degree.

Similarly one can define a disconnected Hurwitz potential $\mathcal{H}^\bullet$ encoding all disconnected Hurwitz numbers.

Fact The connected and disconnected potentials are related by exponentiation:

$$1 + \mathcal{H}^\bullet = e^{\mathcal{H}} \tag{6.2}$$

Example 6.2.3 We have seen in Exercise 6.1 that $H_{0,3} = 4$. From the representation theory:

$$H_{0,3}^\bullet = \frac{1}{36}(2 \cdot 3^4) = \frac{9}{2} = 4 + \frac{1}{2}$$

Looking at the coefficient of $u^4 z q^3$ in Eq. (6.2):

$$H_{0,3}^\bullet \frac{u^4}{4!} z q^3 = H_{0,3} \frac{u^4}{4!} z q^3 + \frac{1}{2!} 2 \left(H_{1,2} \frac{u^4}{4!} q^2 \right) (H_{0,1} z q).$$

Remark 6.2.4 Unfortunately I don't know of any particulary efficient reference for this section. The book [Wil06] contains more information that one might want to start with on generating functions; early papers of various subsets of Goulden, Jackson and Vakil contain the definitions and basic properties of the Hurwitz potential.

6.2.2 Degeneration

Target genus 0, 3-pointed Hurwitz numbers suffice to determine the whole theory of Hurwitz numbers, because of the **degeneration formulas**.

Theorem 6.2.5 (Degeneration Formulas) *Let $\mathfrak{z}(v)$ denote the order of the centralizer of a permutation of cycle type v. Then:*

(1)

$$H_{g \to 0}^{0,\bullet}(\eta_1, \ldots, \eta_s, \mu_1, \ldots, \mu_t) = \sum_{v \vdash d} \frac{\mathfrak{z}(v)}{(\mathrm{Aut}v)^2} H_{g_1 \to 0}^{0,\bullet}$$

$$(\eta_1, \ldots, \eta_s, v) H_{g_2 \to 0}^{0,\bullet}(v, \mu_1, \ldots, \mu_t)$$

with $g_1 + g_2 + \ell(v) - 1 = g$.

(2)

$$H_{g \to 1}^{0,\bullet}(\eta_1, \ldots, \eta_s) = \sum_{v \vdash d} \frac{\mathfrak{z}(v)}{(\mathrm{Aut}v)^2} H_{g - \ell(v) \to 0}^{0,\bullet}(\eta_1, \ldots, \eta_s, v, v).$$

These formulas are called degeneration formulas because geometrically they correspond to simultaneously degenerating the source and the target curve, as illustrated in Fig. 6.3. We will discuss the geometric perspective on degeneration formulas and how

Fig. 6.3 Degeneration of a cover to a nodal cover. Note that source and target degenerate simultaneously and the ramification orders on both sides of the node match

subtle issues of infinitesimal automorphisms (that explain the factor of $\mathfrak{z}(\nu)$) arise. A combinatorial proof is straightforward, and we present it here.

Proof of (1)   Recall that

$$d! H^{0,\bullet}_{g \to 0}(\eta_1, \ldots, \eta_s, \mu_1, \ldots, \mu_t) = |\{\sigma_1, \ldots, \sigma_s, \tilde{\sigma}_1, \ldots, \tilde{\sigma}_t\}|,$$

where the permutations have the appropriate cycle type, and the product of all permutations is the identity. Define $\pi = \sigma_1 \ldots \sigma_s$, then

$$|\{\sigma_1, \ldots, \sigma_s, \pi^{-1}, \pi, \tilde{\sigma}_1, \ldots, \tilde{\sigma}_t\}| = \sum_{\nu \vdash d} \frac{1}{|C_\nu|} |\{\sigma_1, \ldots, \sigma_s, \pi_1\}| |\{\pi_2, \tilde{\sigma}_1, \ldots, \tilde{\sigma}_t\}|$$

where in the RHS π_1 and π_2 have cycle type ν and we require the products of the permutations in the two sets to equal the identity. We must divide by $|C_\nu|$ because in the LHS we want the two newly introduced permutations to be inverses of each other, and not just in the same conjugacy class. But now we recognize that the term on the RHS is:

$$\sum_{\nu \vdash d} \frac{1}{|C_\nu|} d! H^{0,\bullet}_{g_1 \to 0}(\eta_1, \ldots, \eta_s, \nu) d! H^{0,\bullet}_{g_2 \to 0}(\nu, \mu_1, \ldots, \mu_t)$$

The proof is finally concluded by observing the identity $|C_\nu| \mathfrak{z}(\nu) = d!$. $\square$

6.3   Hurwitz Numbers and Moduli Spaces

For the sake of self-containedness, we quickly introduce some families of moduli spaces that are related to Hurwitz theory.

$\overline{\mathcal{M}}_{g,n}$ the moduli space of (isomorphism classes of) **stable curves** of genus g with n marked points. Stability means that every rational component must have at least three special points (nodes or marks), and that a smooth genus one curve needs to have at least one mark. $\overline{\mathcal{M}}_{g,n}$ is a smooth stack of dimension $3g - 3 + n$, connected, irreducible. See [HM98] for more.

$\overline{\mathcal{M}}_{g,n}(\alpha_1, \ldots, \alpha_n)$ in **weighted stable curves** ([Has03]) one tweaks the stability of a pointed curve $(X = \cup_j X_j, p_1, \ldots, p_n)$ by assigning weights α_i to the marked points and requiring the restriction to each X_j of $\omega_X + \sum \alpha_i p_i$ to be ample (this amounts to the combinatorial condition that $\sum_{p_i \in X_j} \alpha_i + n_j > 2 - 2g_j$, where n_j is the number of shadows of nodes on the j-th component of the normalization of X and g_j is the geometric genus of such component). In these spaces "light" points can collide with each other until a "critical mass" is reached that forces the sprouting of new components.

When $g = 0$, two points are given weight 1 and all other points very small weight, the space $\overline{M}_{0,2+r}(1, 1, \varepsilon, \ldots, \varepsilon)$ is classically known as the *Losev-Manin* space [LM00]: it parameterizes chains of $\mathbb{P}^1$'s with the heavy points on the two external components and light points (possibly overlapping amongst themselves) in the smooth locus of the chain.

$\mathcal{M}_{g,n}(X, \beta)$ the space of **stable maps** to X of degree $\beta \in H_2(X)$. A map is stable if every contracted rational component has three special points. If $g = 0$ and X is convex then these are smooth schemes, but in general these are nasty creatures even as stacks. They are singular and typically non-equidimensional. Luckily deformation theory experts can construct a Chow class, called **virtual fundamental class**, of degree in the expected dimension, and enjoying many of the formal properties of the fundamental class. Intersection theory on these spaces is then rescued by capping with the virtual fundamental class. Good references for people interested in these spaces are [HKK+03], [KV07] and [FP97].

$Hurw_{g \to h,d}(\underline{\eta}) \subset Adm_{g \to h,d}(\underline{\eta})$ the **Hurwitz spaces** parameterize degree d covers of smooth curves of genus h by smooth curves of genus g. A vector of partitions of d specifies the ramification profiles over marked points on the base. All other ramification is required to be simple. Hurwitz spaces are typically smooth schemes (unless the ramification profiles are chosen in very particular ways so as to allow automorphisms), but they are obviously non compact. The **admissible cover** compactification, consisting of degenerating simultaneosly target and cover curves, was introduced in [HM82]. In [ACV01], the normalization of such space is interpreted as a (component of a) space of stable maps to the stack BS_d. Without going into the subtleties of stable maps to a stack, we understand that by admissible cover we always denote the corresponding smooth stack.

$\mathcal{M}_{g,n}(X, \beta; \alpha D)$ spaces of **relative stable maps** relative to a divisor D with prescribed tangency conditions([LR01, Li02b]). We are especially interested in the case when X is itself a curve. In this case giving relative conditions is equivalent to specifying ramification profiles over some marked points of the target: spaces of relative stable maps are a "hybrid" compactification that behaves like admissible covers over the relative points and as stable maps elsewhere. See [Vak08] for a more detailed description of the boundary degenerations.

Remark 6.3.1 When the target space is $\mathbb{P}^1$, an important variation of spaces of (relative) stable maps is the so called space of **rubber** maps, or maps to an unparameterized $\mathbb{P}^1$, where two maps are considered equivalent when they agree up to an automorphism of the base $\mathbb{P}^1$ preserving 0 and ∞ (in other words a $\mathbb{C}^*$ scaling of the base). It will be clear later why we care about these spaces.

There are important morphism connecting these types of moduli spaces. If we denote by Cov (some compactification of) the Hurwitz space of covers of $\mathbb{P}^1$ (by admissible covers, stable maps, relative stable maps, ...), we have a natural diagram:

$$
\begin{array}{ccc}
Cov & \xrightarrow{\ s\ } & \overline{\mathcal{M}}_{g,n} \\
\Big\downarrow {\scriptstyle br} & & \\
Tar & &
\end{array}
\tag{6.3}
$$

where Tar denotes a moduli space of branch divisors on the target of the cover, and br the branch morphism that assigns to each cover $f : C \to \mathbb{P}^1$ its branch divisor.

Fact (Important) The Hurwitz number equals the degree of the branch morphism.

6.3.1 Tautological Bundles on Moduli Spaces

We define bundles on moduli spaces by describing them in terms of the geometry of families of objects. In other words, for any family $X \to B$, we give a bundle on B constructed in some canonical way from the family X. This insures that this assignment is compatible with pullbacks (morally means that we are thinking of B as a chart and that the bundle patches along various charts).

The Cotangent Line Bundle and ψ Classes

An excellent reference for this section, albeit unfinished and unpublished, is [Koc01].

Definition 6.3.2 The i-th **cotangent line bundle** $\mathbb{L}_i \to \overline{\mathcal{M}}_{g,n}$ is globally defined as the restriction to the i-th section of the relative dualizing sheaf from the universal family:

$$
\mathbb{L}_i := \sigma_i^*(\omega_\pi).
$$

The first Chern class of the cotangent line bundle is called ψ **class**:

$$
\psi_i := c_1(\mathbb{L}_i).
$$

This definition is slick but unenlightening, so let us chew on it a bit. Given a family of marked curves $f : X \to B (= \varphi_f : B \to \overline{\mathcal{M}}_{g,n})$, the cotangent spaces of the fibers X_b at the i-th mark naturally fit together to define a line bundle on the image of the i-th section, which is then isomorphic to the base B. This line bundle is the pullback $\varphi_f^*(\mathbb{L}_i)$. Therefore informally one says that the cotangent line bundle is the line bundle whose fiber over a moduli point is the cotangent line of the parameterized curve at the i-th mark.

When two moduli spaces admitting ψ classes are related by natural morphisms, a natural question to ask is how the corresponding ψ classes compare (more precisely, how a ψ class in one space compares with the pull-back via the natural morphism of the corresponding ψ class on the other space). The answer is provided by the following Lemma.

Lemma 6.3.3 *The following comparisons of ψ classes hold.*

(1) *Let $\pi_{n+1} : \overline{\mathcal{M}}_{g,n+1} \to \overline{\mathcal{M}}_{g,n}$ be the natural forgetful morphism, and $i \neq n + 1$. Then*

$$\psi_i = \pi_{n+1}^* \psi_i + D_{i,n+1},$$

where $D_{i,n+1}$ is the boundary divisor parameterizing curves where the i-th and the $n + 1$-th mark are the only two marks on a rational tail (or the image of the i-th section, if you think of $\overline{\mathcal{M}}_{g,n+1}$ as the universal family of $\overline{\mathcal{M}}_{g,n}$).
(2) *Let $\pi : \overline{\mathcal{M}}_{g,1}(X, \beta) \to \overline{\mathcal{M}}_{g,1}$ be the natural forgetful morphism. Then*

$$\psi_1 = \pi^* \psi_1 + D_1,$$

where D_1 is the divisor of maps where the mark lies on a contracting rational tail.
(3) *Let $r : \overline{\mathcal{M}}_{g,n}(\alpha_1, \ldots, \alpha_n) \to \overline{\mathcal{M}}_{g,n}(\alpha_1', \ldots, \alpha_n')$ be the natural reduction morphism. Then*

$$\psi_i = r^* \psi_i + D,$$

where D is the boundary divisor parameterizing curves where the i-th mark lies on a component that is contracted in $\overline{\mathcal{M}}_{g,n}(\alpha_1', \ldots, \alpha_n')$.

In all cases the intuitive idea is that the "difference" in the cotangent line bundles is supported on the locus where the mark lives on a curve in the first space that gets contracted in the second space. To make a formal proof one has to observe how the universal family of the first space is obtained by appropriately blowing up the pull-back of the universal family on the second space, and what effect that has on the normal bundle to a section.

The Hodge Bundle

Definition 6.3.4 The **Hodge bundle** $\mathbb{E}(=\mathbb{E}_{g,n})$ is a rank g bundle on $\overline{\mathcal{M}}_{g,n}$, defined as the pushforward of the relative dualizing sheaf from the universal family. Over a curve X, the fiber is canonically $H^0(X, \omega_X)$ (i.e. the vector space of holomorphic 1-forms if X is smooth). The Chern classes of $\mathbb{E}$ are called λ classes:

$$\lambda_i := c_i(\mathbb{E}).$$

We recall the following properties ([Mum83]):

Mumford Relation the total Chern class of the sum of the Hodge bundle with its dual is trivial:

$$c(\mathbb{E} \oplus \mathbb{E}^\vee) = 1. \tag{6.4}$$

Hence $\mathrm{ch}_{2i} = 0$ if $i > 0$.

Separating nodes

$$\iota^*_{g_1,g_2,S}(\mathbb{E}) \cong \mathbb{E}_{g_1,n_1} \oplus \mathbb{E}_{g_2,n_2}, \tag{6.5}$$

where with abuse of notation we omit pulling back via the projection maps from $\overline{\mathcal{M}}_{g_1,n_1+1} \times \overline{\mathcal{M}}_{g_2,n_2+1}$ onto the factors.

Non-separating nodes

$$\iota^*_{irr}(\mathbb{E}) \cong \mathbb{E}_{g-1,n} \oplus \mathcal{O}. \tag{6.6}$$

Remark 6.3.5 We define the Hodge bundle and λ classes on moduli spaces of stable maps and Hurwitz spaces by pulling back via the appropriate forgetful morphisms.

6.4   Simple Hurwitz Numbers and the ELSV Formula

The name **simple Hurwitz number** (denoted $H_g(\eta)$) is reserved for Hurwitz numbers to a base curve of genus 0, and with only one special point where arbitrary ramification is assigned. In this case the number of simple ramification, determined by the Riemann-Hurwitz formula, is

$$r = 2g + d - 2 + \ell(\eta). \tag{6.7}$$

Definition 6.2.1 simplifies further to count (up to an appropriate multiplicative factor) the number of ways to factor a (fixed) permutation $\sigma \in C_\eta$ into r transpositions that generate S_d:

$$H_g(\eta) = \frac{1}{\prod \eta_i} |\{(\tau_1, \ldots, \tau_r \ s.t. \ \tau_1 \cdot \ldots \cdot \tau_r = \sigma \in C_\eta, \langle \tau_1, \ldots, \tau_r \rangle = S_d)\}| \qquad (6.8)$$

The first formula for simple Hurwitz number was given and "sort of" proved by Hurwitz in 1891 ([Hur91]):

$$H_0(\eta) = r! d^{\ell(\eta)-3} \prod \frac{\eta_i^{\eta_i}}{\eta_i!}.$$

Particular cases of this formula were proved throughout the last century, and finally the formula became a theorem in 1997 ([GJ97]). In studying the problem for higher genus, Goulden and Jackson made the following conjecture.

Conjecture (Goulden-Jackson Polynomiality Conjecture) For any fixed values of $g, n :=$ $\ell(\eta)$:

$$H_g(\eta) = r! \prod \frac{\eta_i^{\eta_i}}{\eta_i!} P_{g,n}(\eta_1 \ldots, \eta_n), \qquad (6.9)$$

where $P_{g,n}$ is a symmetric polynomial in the η_i's with:

- deg $P_{g,n} = 3g - 3 + n$;
- $P_{g,n}$ doesn't have any term of degree less than $2g - 3 + n$;
- the sign of the coefficient of a monomial of degree d is $(-1)^{d-(3g+n-3)}$.

In [ELSV01] Ekehdal, Lando, Shapiro and Vainshtein prove this formula by establishing a remarkable connection between simple Hurwitz numbers and tautological intersections on the moduli space of curves.

Theorem 6.4.1 (ELSV Formula) *For all values of* $g, n = \ell(\eta)$ *for which the moduli space* $\overline{\mathcal{M}}_{g,n}$ *exists:*

$$H_g(\eta) = r! \prod \frac{\eta_i^{\eta_i}}{\eta_i!} \int_{\overline{\mathcal{M}}_{g,n}} \frac{1 - \lambda_1 + \ldots + (-1)^g \lambda_g}{\prod(1 - \eta_i \psi_i)}, \qquad (6.10)$$

Remark 6.4.2 Goulden and Jackson's polynomiality conjecture is proved by showing the coefficients of $P_{g,n}$ as tautological intersection numbers on $\overline{\mathcal{M}}_{g,n}$. Using standard multi-index notation:

$$P_{g,n} = \sum_{k=0}^{g} \sum_{|I_k|=3g-3+n-k} (-1)^k \left(\int \lambda_k \psi^{I_k} \right) \eta^{I_k}$$

Remark 6.4.3 The polynomial $P_{g,n}$ is a generating function for all linear (meaning where each monomial has only one λ class) Hodge integrals on $\overline{\mathcal{M}}_{g,n}$, and hence a good understanding of this polynomial can yield results about intersection theory on the moduli spaces of curves. In fact the *ELSV* formula has given rise to several remarkable applications:

[OP09] Okounkov and Pandharipande use the ELSV formula to give a proof of Witten's conjecture, that an appropriate generating function for the ψ intersections satisfies the KdV hierarchy. The ψ intersections are the coefficients of the leading terms of $P_{g,n}$, and hence can be reached by studying the asymptotics of Hurwitz numbers:

$$\lim_{N \to \infty} \frac{P_{g,n}(N\eta)}{N^{3g-3+n}}$$

[GJV06] Goulden, Jackson and Vakil get a handle on the lowest order terms of $P_{g,n}$ to give a new proof of the λ_g conjecture:

$$\int_{\overline{\mathcal{M}}_{g,n}} \lambda_g \psi^I = \binom{2g-3+n}{I} \int_{\overline{\mathcal{M}}_{g,1}} \lambda_g \psi_1^{2g-2}$$

We sketch a proof of the *ELSV* formula following [GV03]. The strategy is to evaluate an integral via localization, fine tuning the geometry in order to obtain the desired result.

Denote:

$$\mathcal{M} := \overline{\mathcal{M}}_g(\mathbb{P}^1, \eta\infty)$$

the moduli space of relative stable maps of degree d to $\mathbb{P}^1$, with profile η over ∞. The degenerations included to compactify are twofold:

- away from the preimages of ∞ we have degenerations of "stable maps" type: we can have nodes and contracting components for the source curve, and nothing happens to the target $\mathbb{P}^1$;
- when things collide at ∞, then the degeneration is of "admissible cover" type: a new rational component sprouts from $\infty \in \mathbb{P}^1$, the special point carrying the profile

requirement transfers to this component. Over the node we have nodes for the source curve, with maps satisfying the kissing condition.

The space $\mathcal{M}$ has virtual dimension $r = 2g + d + \ell(\eta) - 2$ and admits a globally defined branch morphism ([FP02]):

$$br : \mathcal{M} \rightarrow Sym^r(\mathbb{P}^1) \cong \mathbb{P}^r.$$

The simple Hurwitz number:

$$H_g(\eta) = \deg(br) = br^*(pt.) \cap [\mathcal{M}]^{vir}$$

can now interpreted as an intersection number on a moduli space with a torus action and evaluated via localization. The map br can be made $\mathbb{C}^*$ equivariant by inducing the appropriate action on $\mathbb{P}^r$. The key point is now to choose the appropriate equivariant lift of the class of a point in $\mathbb{P}^r$. Recalling that choosing a point in $\mathbb{P}^r$ is equivalent to fixing a branch divisor, we choose the $\mathbb{C}^*$ fixed point corresponding to stacking all ramification over 0. Then there is a unique fixed locus contributing to the localization formula, depicted in Fig. 6.4, which is essentially isomorphic to $\overline{\mathcal{M}}_{g,n}$ (up to some automorphism factors coming from the bubbles over $\mathbb{P}^1$).

The *ELSV* formula falls immediately out of the localization formula. The virtual normal bundle to the unique contributing fixed locus has a denominator part given from the smoothing of the nodes that produces the denominator with ψ classes in the ELSV formula. Then there is the equivariant Euler class of the derived push-pull of $T\mathbb{P}^1(-\infty)$: when restricted to the fixed locus this gives a Hodge bundle linearized with weight

Fig. 6.4 the unique contributing fixed locus in the localization computation proving the *ELSV* formula

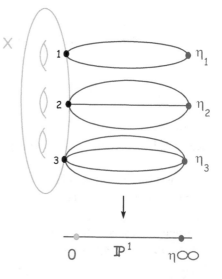

1, producing the polynomial in λ classes, and a bunch of trivial but not equivariantly trivial bundles corresponding to the restriction of the push-pull to the trivial covers of the main components. The equivariant Euler class of such bundles is just the product of the corresponding weights, and gives rise to the combinatorial pre-factors before the Hodge integral.

Remark 6.4.4 An abelian orbifold version of the ELSV formula has been developed by Johnson, Pandharipande and Tseng in [JPT11]. In this case the connection is made between Hurwitz-Hodge integrals and wreath Hurwitz numbers.

6.5   Appendix: Atyiah-Bott Localization

This section is meant as a friendly introduction to localization for people that may not have encountered it before. It does not contain sufficient information for a person to be able to use this technique, but the intention is that it may make the references that do (e.g. [HKK+03]) become significantly more approachable.

The localization theorem of [AB84] is a powerful tool for the intersection theory of moduli spaces that can be endowed with a torus action.

6.5.1 Equivariant Cohomology

Let G be a group acting on a space X. According to your point of view G might be a compact Lie group or a reductive algebraic group. Then G-equivariant cohomology is a cohomology theory developed to generalize the notion of the cohomology of a quotient when the action of the group is not free. The idea is simple: since cohomology is homotopy invariant, replace X by a homotopy equivalent space $\tilde{X}$ on which G acts freely, and then take the cohomology of $\tilde{X}/G$. Rather than delving into the definitions that can be found in [HKK+03], Chapter 4, we recall here some fundamental properties that we use:

1. If G acts freely on X, then

$$H_G^*(X) = H^*(X/G).$$

2. If X is a point, then let EG be any contractible space on which G acts freely, $BG := EG/G$, and define:

$$H_G^*(pt.) = H^*(BG).$$

3. If G acts trivially on X, then

$$H_G^*(X) = H^*(X) \otimes H^*(BG).$$

Example 6.5.1 If $G = \mathbb{C}^*$, then $EG = S^\infty$, $BG := \mathbb{P}^\infty$ and

$$H_{\mathbb{C}^*}^*(pt.) = \mathbb{C}[\hbar],$$

with $\hbar = c_1(\mathcal{O}(1))$.

Remark 6.5.2 Dealing with infinite dimensional spaces in algebraic geometry is iffy. In [Ful98], Fulton finds an elegant way out by showing that for any particular degree of cohomology one is interested in, one can work with a finite dimensional approximation of BG. Another route is to instead work with the stack $\mathcal{B}G = [pt./G]$. Of course the price to pay is having to formalize cohomology on stacks... here let us just say that $\mathcal{O}(1) \to \mathcal{B}\mathbb{C}^*$, pulled back to the class of a point, is a copy of the identity representation $Id : \mathbb{C}^* \to \mathbb{C}^*$.

Let $\mathbb{C}^*$ act on X and let F_i be the irreducible components of the fixed locus. If we-push forward and then pull-back the fundamental class of F_i we obtain

$$i^* i_*(F_i) = e(N_{F_i/X}).$$

Since $N_{F_i/X}$ is the moving part of the tangent bundle to F_i, this Euler class is a polynomial in $\hbar$ where the $\hbar^{codim(F_i)}$ term has non-zero coefficient. This means that if we allow ourselves to invert $\hbar$, this Euler class becomes invertible. This observation is pretty much the key to the following theorem:

Theorem 6.5.3 (Atyiah-Bott Localization) *The maps:*

$$\bigoplus_i H^*(F_i)(\hbar) \xrightarrow{\sum \frac{i_*}{e(N_i)}} H_{\mathbb{C}^*}^*(X) \otimes \mathbb{C}(\hbar) \xrightarrow{i^*} \bigoplus_i H^*(F_i)(\hbar)$$

are inverses (as $\mathbb{C}(\hbar)$-algebra homomorphisms) of each other. In particular, since the constant map to a point factors (equivariantly!) through the fixed loci, for any equivariant cohomology class α:

$$\int_X \alpha = \sum_i \int_{F_i} \frac{i^*(\alpha)}{e(N_{F_i/X})}$$

In practice, one can reduce the problem of integrating classes on a space X, which might be geometrically complicated, to integrating over the fixed loci (which are hopefully simpler).

Example 6.5.4 (The Case of $\mathbb{P}^1$) Let $\mathbb{C}^*$ act on a two dimensional vector space V by:

$$t \cdot (v_0, v_1) := (v_0, tv_1)$$

This action defines an action on the projectivization $\mathbb{P}(V) = \mathbb{P}^1$. The fixed points for the torus action are $0 = (1 : 0)$ and $\infty = (0 : 1)$. The canonical action on $T_{\mathbb{P}}$ has weights $+1$ at 0 and -1 at ∞. Identifying $V - 0$ with the total space of $\mathcal{O}_{\mathbb{P}^1}(-1)$ minus the zero section, we get a canonical lift of the torus action to $\mathcal{O}_{\mathbb{P}^1}(-1)$, with weights $0, 1$. Also, since $\mathcal{O}_{\mathbb{P}^1}(1) = \mathcal{O}_{\mathbb{P}^1}(-1)^\vee$, we get a natural linearization for $\mathcal{O}_{\mathbb{P}^1}(1)$ as well (with weights $0, -1$). Finally, by thinking of $\mathbb{P}^1$ as the projectivization of an equivariant bundle over a point, we obtain:

$$H^*_{\mathbb{C}^*}(\mathbb{P}^1) = \frac{\mathbb{C}[H, \hbar]}{H(H - \hbar)}.$$

The Atyiah-Bott isomorphism now reads:

$$\mathbb{C}(\hbar)_0 \oplus \mathbb{C}(\hbar)_\infty \leftrightarrow H^*_{\mathbb{C}^*}(\mathbb{P}) \otimes \mathbb{C}(\hbar)$$

$(1, 0)$	$\rightarrow$	$\frac{H}{\hbar}$
$(0, 1)$	$\rightarrow$	$\frac{H-\hbar}{-\hbar}$
$(1, 1)$	$\leftarrow$	1
$(\hbar, 0)$	$\leftarrow$	H

Applying the Localization Theorem to Spaces of Maps

Kontsevich first applied the localization theorem to smooth moduli spaces of maps in [Kon95]. Graber and Pandharipande [GP99] generalized this technique to the general case of singular moduli spaces, showing that localization "plays well" with the virtual fundamental class.

Let X be a space with a $\mathbb{C}^*$ action, admitting a finite number of fixed points P_i, and of fixed lines l_i (NOT pointwise fixed). Typical examples are given by projective spaces, flag varieties, toric varieties... Then:

1. A $\mathbb{C}^*$ action is naturally induced on $\overline{\mathcal{M}}_{g,n}(X, \beta)$ by postcomposition.
2. The fixed loci in $\overline{\mathcal{M}}_{g,n}(X, \beta)$ parameterize maps from nodal curves to the target such that:

   - components of arbitrary genus are contracted to the fixed points P_i.
   - rational components are mapped to the fixed lines as d-fold covers fully ramified over the fixed points.

In particular

$$F_i \cong \prod \overline{\mathcal{M}}_{g_j,n_j} \times \prod \mathcal{B}\mathbb{Z}_{d_k}.$$

3. The **"virtual"** normal directions to the fixed loci correspond essentially to either smoothing the nodes of the source curve (which by exercise 6.10 produces sums of ψ classes and equivariant weights), or to deforming the map out of the fixed points and lines. This can be computed using the deformation exact sequence ([HKK+03], (24.2)), and produces a combination of equivariant weights and λ classes.

The punchline is, one has reduced the tautological intersection theory of $\overline{\mathcal{M}}_{g,n}(X, \beta)$ to combinatorics, and Hodge integrals (i.e. intersection theory of λ and ψ classes). From a combinatorial point of view this can be an extremely complicated and often unmanageable problem, but in principle application of the Grothendieck-Riemann-Roch Theorem and of Witten Conjecture/Kontsevich's Theorem completely determine all Hodge integrals. Carel Faber in [Fab99] explained this strategy and wrote a Maple code that can handle efficiently integrals up to a certain genus and number of marks.

6.6   Exercises

Exercise 6.1 Use Definition 6.2.1 to check the Hurwitz numbers in Example 6.1.2. Compute $H^4_{1\to 0,3}((3)) = 9$ and $H^4_{0\to 0,3} = 4$.

Exercise 6.2 Convince yourself of Eq. (6.2), relating the connected and disconnected Hurwitz potentials. To me, this is one of those mathematical facts that are mysterious until you stare at it long enough that, all of a sudden, it becomes obvious. . .

Exercise 6.3 Check the coefficients of Eq. (6.2) for the monomials corresponding to the disconnected Hurwitz numbers $H^\bullet_{-1,4}$, $H^\bullet_{-1}((2, 1, 1), (2, 1, 1))$ and $H^\bullet_{-1}((2, 1, 1), (2, 1, 1), (2, 1, 1), (2, 1, 1))$. All these Hurwitz numbers are equal to $\frac{3}{4}$.

Exercise 6.4 Figure out a combinatorial formula for $\mathfrak{z}(\nu)$, and interpret it as the size of the automoprhism group of a cover of $\mathbb{P}^1$ by a bunch of $\mathbb{P}^1$'s with only two branch points.

Exercise 6.5 Prove part (2) of Theorem 6.2.5, giving the degeneration formula for Hurwitz numbers when a non-separating loop is contracted.

Exercise 6.6 Describe the moduli space $M_g(\mathbb{P}^1, 1)$ and the stable maps compactification $\overline{M}_g(\mathbb{P}^1, 1)$.

Exercise 6.7 The **hyperelliptic locus** is the subspace of $\overline{M}_g$ parameterizing curves that admit a double cover to $\mathbb{P}^1$. Understand the hyperelliptic locus as the moduli space $Adm_{g\to 0,2}((2), \ldots, (2))$ and subsequently as a stack quotient of $\overline{\mathcal{M}}_{0,2g+2}$ by the trivial action of $\mathbb{Z}_2$.

Exercise 6.8 Referring to the universal diagram (6.3), understand how Tar depends on the choice of Cov. In particular, figure out what it is when Cov equals the Hurwitz space, the admissible cover compactification, the stable maps compactification, the compactification by relative stable maps.

Exercise 6.9 Convince yourself that the normal bundle to the image of the i-th section in the universal family of the moduli space of curves is naturally isomorphic to $\mathbb{L}_i^\vee$ (This is sometimes called the i-th tangent line bundle and denoted $\mathbb{T}_i$).

Exercise 6.10 Consider an irreducible boundary divisor $D \cong \overline{M}_{g_1,n_1+\bullet} \times \overline{M}_{g_2,n_2+\star}$. Then the normal bundle of D in the moduli space is naturally isomorphic to the tensor product of the tangent line bundles of the components at the shadows of the node:

$$N_{D/\overline{M}_{g,n}} \cong \mathbb{L}_\bullet^\vee \boxtimes \mathbb{L}_\star^\vee$$

Is this statement consistent with the previous exercise? Why?

Exercise 6.11 Show that Lemma 6.3.3 gives sufficient information to determine ψ classes for every $\overline{M}_{0,n}$. In particular show it gives the following useful combinatorial boundary description of a ψ class. Let i, j, k be three distinct marks. The class ψ_i is the sum of all boundary divisors parameterizing curves where the i-th mark is on one component, the j-th and k-th marks are on the other. Note that such a boundary description is not unique, as it depends on the choice of j and k.

Exercise 6.12 Use the properties (6.4), (6.5) and (6.6) of the Hodge bundle, together with the formal properties of Chern classes, to show vanishing properties of λ-classes:

1. $\lambda_g^2 = 0$ if $g > 0$.
2. $\lambda_g \lambda_{g-1}$ vanishes on the boundary of $\overline{M}_g$. If now we allow marked points, then the vanishing holds on "almost all" the boundary, but one needs to be more careful. Describe the vanishing locus of $\lambda_g \lambda_{g-1}$ in this case.
3. λ_g vanishes on the locus of curves not of compact type (i.e. where the geometric and arithmetic genera are different).

Tropical Hurwitz Theory

7

In this lecture we explore how the degeneration formulas make a connection between Hurwitz theory and tropical geometry, and how such a perspective sheds light on the remarkable combinatorial properties of double Hurwitz numbers.

7.1 Double Hurwitz Numbers

Double Hurwitz Numbers count covers of $\mathbb{P}^1$ with special ramification profiles over two points, that for simplicity we assume to be 0 and ∞. Double Hurwitz numbers are denoted $H_g^r(\mathbf{x})$, for $\mathbf{x} \in H \subset \mathbb{R}^n$ an integer lattice point on the hyperplane $\sum x_i = 0$. The subset of positive coordinates corresponds to the profile over 0 and the negative coordinates to the profile over ∞. We define $\mathbf{x_0} := \{x_i > 0\}$ and $\mathbf{x_\infty} := \{x_i < 0\}$.

The number r of simple ramification is given by the Riemann-Hurwitz formula,

$$r = 2g - 2 + n$$

and it is independent of the degree d. In [GJV03], Goulden, Jackson and Vakil start a systematic study of double Hurwitz numbers and in particular invite us to consider them as a function:

$$H_g^r(-) : \mathbb{Z}^n \cap H \to \mathbb{Q}. \tag{7.1}$$

They prove some remarkable combinatorial property of this function:

Theorem 7.1.1 (GJV) *The function $H_g(-)$ is a piecewise polynomial function of degree $4g - 3 + n$.*

© The Author(s), under exclusive license to Springer Nature Switzerland AG 2023
R. Cavalieri et al., *Tropical and Logarithmic Methods in Enumerative Geometry*,
Oberwolfach Seminars 52, https://doi.org/10.1007/978-3-031-39401-0_7

And conjecture some more:

Conjecture (GJV) The polynomials describing $H_g^r(-)$ have degree $4g - 3 + n$, lower degree bounded by $2g - 3 + n$ and are even or odd polynomials (depending on the parity of the leading coefficient).

Shadrin et al. [SSV08] describe the chambers of polynomiality and give wall-crossing formulas for double Hurwitz numbers in genus 0. Their results are generalized to arbitrary genus in [CJM11]. Tropical geometry gives an approach to the study of double Hurwitz numbers that shows the conceptual reason for these combinatorial structure results.

7.2   Tropical Double Hurwitz Numbers

Tropical Hurwitz numbers arise as the degree of a tropical branch morphism among appropriate moduli spaces of covers of tropical curves. Such degree is computed by a combinatorial formula in terms of appropriately decorated graphs.

Definition 7.2.1 For fixed g and $\mathbf{x} = (x_1, \ldots, x_n)$, a graph Γ is a **monodromy graph** if:

(1) Γ is a connected, genus g, directed graph.
(2) Γ has n 1-valent vertices called *leaves*; the edges leading to them are *ends*. All ends are directed inward, and are labeled by the weights $x_1, \ldots, x_n$. If $x_i > 0$, we say it is an *in-end*, otherwise it is an *out-end*.
(3) All other vertices of Γ are 3-valent, and are called *internal vertices*. Edges that are not ends are called *internal edges*.
(4) After reversing the orientation of the out-ends, Γ does not have directed loops, sinks or sources.
(5) The internal vertices are ordered compatibly with the partial ordering induced by the directions of the edges.
(6) Every internal edge e of the graph is equipped with a *weight* $w(e) \in \mathbb{N}$. The weights satisfy the *balancing condition* at each internal vertex: the sum of all weights of incoming edges equals the sum of the weights of all outgoing edges.

Using monodromy graphs, tropical Hurwitz numbers are defined in [CJM10].

Definition 7.2.2 The tropical (double) Hurwitz number $H_g^{trop}(\mathbf{x})$ is defined as:

$$H_g^{trop}(\mathbf{x}) = \sum_{\Gamma} \frac{1}{|Aut(\Gamma)|} \varphi_{\Gamma}, \tag{7.2}$$

where the sum is over all monodromy graphs Γ for g and $\mathbf{x}$, and φ_Γ denotes the product of weights of all internal edges.

Example 7.2.3 Here are some examples of tropical Hurwitz numbers.

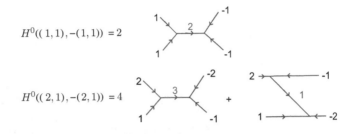

$$H^0((1,1),-(1,1)) = 2$$

$$H^0((2,1),-(2,1)) = 4$$

The definition of tropical Hurwitz numbers is motivated by tropical intersection theory. Monodromy graphs are an example of tropical covers of the tropical projective line. One may consider a tropical moduli space[1] of maps to tropical $\mathbb{P}^1$, denoted $M^{trop}(\mathbb{P}^1, \mathbf{x})$. There is a natural branch morphism

$$br : M^{trop}(\mathbb{P}, \mathbf{x}) \to \mathbb{R}_{\geq 0}^{r-1}, \qquad (7.3)$$

and a tropical Hurwitz number is defined to be its degree. The tropical branch morphism is a map of equidimensional cone complexes. Its degree may be therefore computed as follows: the local degree at a point inside a maximal dimensional cone σ_F equals to the lattice index of the image of the integral lattice of σ_F inside the integral lattice of $br(\sigma_F)$.

Figure 7.1 illustrates how piecewise polynomiality and wall crossings naturally arise for tropical Hurwitz numbers. Local polynomiality arises from the balancing condition: the weight of the egdes of monodromy graphs are linear homogeneous polynomials in the x_i's (in higher genus there are g additional variables that need to be integrated over the lattice points of a g-dimensional polytope), showing that each graphs contributes with a polynomial multiplicity or the correct degree. As one may see in the example, different graphs contribute according to the sign of $x_1 + y_1$, giving rise to two different polynomials function.

---

[1] In [CJM10] we don't allow tropical maps to contract any part of the source tropical curve. This should be thought as the tropicalization of the closure of the main component of the space of rubber relative stable maps.

Fig. 7.1 Computing double
Hurwitz numbers using
Definition 7.2.2 and observing
the wall crossing

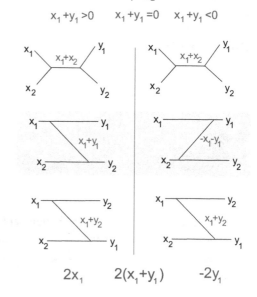

Tropical Hurwitz numbers are related to algebraic ones via a correspondence theorem.

Theorem 7.2.4 (Theorem 5.28 [CJM10])

$$H_g^{trop}(x) = H_g(x) \tag{7.4}$$

We now look at a couple different ways to understand this theorem.

7.3   Correspondence by Cut and Join

The *Cut and Join equations* are a collection of recursions among Hurwitz numbers. In the most elegant and powerful formulation they are expressed as one differential operator acting on the Hurwitz potential. Here we limit ourselves to a basic discussion, and refer the reader to [GJ99] for a more in-depth presentation.

Let $\sigma \in S_d$ be a fixed element of cycle type $\eta = (n_1, \ldots, n_l)$, written as a composition of disjoint cycles as $\sigma = c_l \ldots c_1$. Let $\tau = (ij) \in S_d$ vary among all transpositions. The cycle types of the composite elements $\tau\sigma$ are described below.

cut   if i, j belong to the same cycle (say c_l), then this cycle gets "cut in two": $\tau\sigma$ has cycle type $\eta' = (n_1, \ldots, n_{l-1}, m', m'')$, with $m' + m'' = n_l$. If $m' \neq m''$, there are n_l transpositions giving rise to an element of cycle type η'. If $m' = m'' = n_l/2$, then there are $n_l/2$.

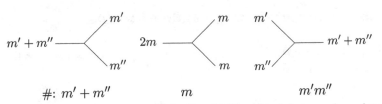

$$\#: m' + m'' \qquad\qquad m \qquad\qquad m'm''$$

Fig. 7.2 Composing with a transposition in S_d. How it effects the cycle type of σ and multiplicity

join   if i, j belong to different cycles (say c_{l-1} and c_l), then these cycles are "joined": $\tau\sigma$ has cycle type $\eta' = (n_1, \dots, n_{l-1} + n_l)$. There are $n_{l-1}n_l$ transpositions giving rise to cycle type η'.

Figure 7.2 illustrates the above discussion.

Example 7.3.1 Let $d = 4$. There are 6 transpositions in S_4. If $\sigma = (12)(34)$ is of cycle type $(2, 2)$, then there are 2 transpositions $((12)$ and (34)) that "cut" σ to give rise to a transposition and $2 \cdot 2$ transpositions $((13), (14), (23), (24))$ that "join" σ into a four-cycle.

To understand how cut and join determines the correspondence theorem, we specialize the definition of Hurwitz number by counting monodromy representations to the case of double Hurwitz numbers.

$$H_g^r(\mathbf{x}) := \frac{|\mathrm{Aut}(\mathbf{x_0})||\mathrm{Aut}(\mathbf{x_\infty})|}{d!} |\{\sigma_0, \tau_1, \dots, \tau_r, \sigma_\infty \in S_d\}|$$

such that:

- σ_0 has cycle type $\mathbf{x_0}$;
- τ_i's are simple transpositions;
- σ_∞ has cycle type $\mathbf{x_\infty}$;
- $\sigma_0\tau_1 \dots \tau_r\sigma_\infty = 1$
- the subgroup generated by such elements acts transitively on the set $\{1, \dots, d\}$.

The key insight is that one can organize this count in terms of the cycle types of the composite elements

$$C\mathbf{x_0} \ni \sigma_0, \sigma_0\tau_1, \sigma_0\tau_1\tau_2, \dots, \sigma_0\tau_1\tau_2 \dots \tau_{r-1}, \sigma_0\tau_1\tau_2 \dots \tau_{r-1}\tau_r \in C\mathbf{x_\infty}$$

At each step the cycle type can change as prescribed by the cut and join recursions, and as in Fig. 7.2 such change may be tracked diagrammatically; for each possibility we can construct a graph with edges weighted by the multiplicites of the cut and join equation. Such graphs are precisely the monodromy graphs and the cut and join multiplicities agree with the those given in Definition 7.2.2.

7.4   Correspondence by Degeneration Formula

To understand how degeneration formula gives rise to the correspondence theorem, it is useful to refer to the diagram:

$$\mathcal{M}(\mathbf{x}) = \overline{\mathcal{M}_g}(\mathbf{x}) \xrightarrow{\quad stab \quad} \overline{\mathcal{M}}_{g,n} \ ,$$

$$\downarrow br$$

$$\mathcal{M}_{br} = \overline{\mathcal{M}}_{0,2+r}(1, 1, \varepsilon, \dots, \varepsilon)/S_r \tag{7.5}$$

and recall that the double Hurwitz number is the degree of the branch morphism br. The degree of $br^*([pt.])$ may be computed by choosing a zero-dimensional boundary stratum $\Delta \in LM(r)$ as a representative for the class of a point. This consists of a chain of r projective lines, with the two branch points with ramification profiles $\mathbf{x}^{\pm}$ on opposite external components of the chain, and exactly one simple branch point on each component of the chain. For any inverse image $(f : C \to T) \in br^{-1}(\Delta)$, the irreducible components of C are rational and contain either two or three special points. In this context, a special point is either a node or a relative point. The degree of $br^*([pt.])$ is then obtained by counting each inverse image $(f : C \to T)$ with the multiplicity prescribed by the *degeneration formula* [LR01, Li02b].

The dual graphs of the source curves of maps $(f : C \to T) \in br^{-1}(\Delta)$ are naturally identified with combinatorial types of tropical covers $F : \Gamma \to \mathbb{R}$ of the tropical line, where the expansion factors of the edges correspond to the ramification orders of the corresponding (shadows of) nodes. This identification gives a bijection between the points $(f : C \to T) \in br^{-1}(\Delta)$ and the *monodromy graphs* from Definition 7.2.1. The correspondence theorem between algebraic and tropical Hurwitz numbers follows from the fact that the local degree of the tropical branch morphism at σ_F equals the degeneration formula multiplicity for the corresponding algebraic cover $f : C \to T$, as described in Exercise 7.4.

7.5   Tropical General Hurwitz Numbers

The correspondence theorem between classical and tropical Hurwitz numbers extends beyond the case of double Hurwitz numbers to any kind of Hurwitz numbers; the general case illustrates an important phenomenon: there are certain parts of the algebraic geometry of covers of curves that are simply not visible to tropical geometry, and must be added to tropical multiplicities in order to obtain correpondence theorems. Sometime this information is referred to as **geometric seed data**. For Hurwitz theory, the geometric seed data consists precisely of target genus zero, three-point Hurwitz numbers.

Fix a vector of partitions $\vec{\mu} = (\mu^1, \ldots, \mu^r)$ of an integer $d > 0$. We wish to study covers of genus g tropical curves, with prescribed ramification data over r points and simple ramification over the remaining s points.

A map of tropical curves satisfies the *local Riemann–Hurwitz condition* if, when $v' \mapsto v$ with local degree d, then

$$2 - 2g(v') = d(2 - 2g(v)) - \sum(m_{e'} - 1), \qquad (7.6)$$

where e' ranges over edges incident to v', and $m_{e'}$ is the expansion factor of the morphism along e'.

Definition 7.5.1 A *tropical admissible cover* of a tropical curve is a harmonic map of tropical curves that satisfies the local Riemann–Hurwitz condition at every point.

Let $\mathcal{H}^{trop}_{g \to h, d}(\vec{\mu})$ denote the space of tropical admissible covers of genus h tropical curves by genus g tropical curves, with expansion factors along infinite edges prescribed by $\vec{\mu}$. It also comes with a tautological branch morphism to $\mathcal{M}^{trop}_{h,r+s}$. The degree of the branch morphism is again used to define tropical Hurwitz numbers.

Definition 7.5.2 Let Θ be a combinatorial type of a tropical admissible cover. We define its weight $\omega(\Theta)$ as the product of:

(1) A factor of $\frac{1}{|Aut_0(\Theta)|}$.
(2) A factor of local Hurwitz numbers $\prod_{v \in \Gamma_{tgt}} H(v)$.
(3) A factor of $M = \prod_{e \in E(\Gamma_{tgt})} M_e$, where M_e is the product of the expansion factors above the edge e.

Remark 7.5.3 We briefly discuss how these factors arise. While (1)–(3) are defined in terms of combinatorics of the tropical covers, they have natural counterparts in the classical theory of admissible covers. Weight (1) accounts for automorphisms of covers lifting the identity map on the target curve. The term (2) encodes the fact that there may be multiple zero dimensional strata in $\overline{\mathcal{H}}_{g,d}(\vec{\mu})$ which have the same dual graph. Finally (3) can be thought of either as "ghost automorphisms" coming from the orbifold structure on a twisted cover [ACV01].

Definition 7.5.4 Let σ_Γ be any fixed top dimensional cone of the tropical moduli space $\mathcal{M}^{trop}_{h,r+s}$. Denote by $\sigma_\Theta^{\mathcal{H}} \mapsto \sigma_\Gamma$ a cone in the moduli space $\mathcal{H}^{trop}_{g,d}(\vec{\mu})$ of combinatorial type Θ such that the base graph of Θ is equal to Γ. The restriction of the tropical branch map is a surjective morphism of cones with integral structure of the same dimension.

Then the **tropical Hurwitz number** is equal to:

$$H_{g\to h,d}(\vec{\mu}) = \sum_{\sigma_{\Theta}^{\mathcal{H}} \mapsto \sigma_{\Gamma}} \omega(\Theta). \tag{7.7}$$

Theorem 7.5.5 ([BBM11]) *Classical and tropical Hurwitz numbers coincide, i.e. we have*

$$H_{g\to h,d}(\vec{\mu}) = H_{g\to h,d}^{trop}(\vec{\mu}).$$

In [CMR16], this theorem is viewed as a consequence of functorial tropicalization. Denote by $\overline{\mathcal{H}}^{an}_{g\to h,d}(\vec{\mu})$ (resp. $\overline{\mathcal{M}}^{an}_{g,n}$) the Berkovich analytification of the space of admissible covers (resp. the moduli space of stable curves).

Theorem 7.5.6 ([CMR16]) *The set theoretic tropicalization map* $trop : \mathcal{H}^{an}_{g\to h,d}(\vec{\mu}) \to \mathcal{H}^{trop}_{g\to h,d}(\vec{\mu})$ *factors through the canonical projection from the analytification to its skeleton* $\Sigma(\overline{\mathcal{H}}^{an}_{g\to h,d}(\vec{\mu}))$,

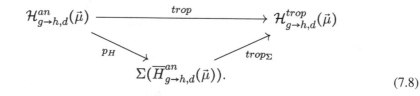

$$(7.8)$$

Furthermore the map $trop_{\Sigma}$ *is a surjective face morphism of cone complexes, i.e. the restriction of* $trop_{\Sigma}$ *to any cone of* $\Sigma(\overline{\mathcal{H}}^{an}_{g\to h,d}(\vec{\mu}))$ *is an isomorphism onto a cone of the tropical moduli space* $\mathcal{H}^{trop}_{g\to h,d}(\vec{\mu})$. *The map* $trop_{\Sigma}$ *extends naturally and uniquely to the extended complexes* $\overline{\Sigma}(\overline{\mathcal{H}}^{an}_{g\to h,d}(\vec{\mu})) \to \overline{\mathcal{H}}^{trop}_{g\to h,d}(\vec{\mu})$.

The map *trop* depends on the choice of the admissible cover compactification even when restricted to the analytification of the Hurwitz space. Intuitively, one may think of a point in $\mathcal{H}^{an}_{g\to h,d}(\vec{\mu})$ as a family of smooth covers over a punctured disk. The tropicalization of such point is obtained by extending the family to an admissible cover and metrizing the dual graph of the central fibers by the valuations of the smoothing parameters of the nodes.

Theorem 7.5.7 ([CMR16]) *Let br denote the branch map* $\overline{\mathcal{H}}_{g\to h,d}(\vec{\mu}) \to \overline{\mathcal{M}}_{h,r+s}$, *and src denote the source map* $\overline{\mathcal{H}}_{g\to h,d}(\vec{\mu}) \to \overline{\mathcal{M}}_{g,n}$, *where n is the number of smooth points in the inverse image of the branch locus. Then the following diagram is commutative:*

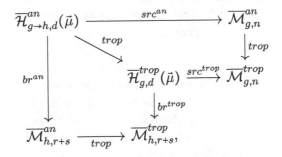

The induced map on skeleta of the branch (resp. source) morphism factors as a composition of the map $trop_\Sigma$ to $\Sigma(\overline{\mathcal{H}}^{an}_{g\to h,d}(\vec{\mu}))$, followed by the tropical branch (resp. source) map, so $br^{trop} = trop_\Sigma \circ br^\Sigma$ (resp. $src^{trop} = trop_\Sigma \circ src^\Sigma$).

One then checks with a local computation that the choice of weights for the cones of the moduli space of tropical admissible covers makes the degree of br^{trop} and br^{an} agree.

7.6 Conjectural ELSV for Double Hurwitz numbers

The combinatorial structure of double Hurwitz numbers seems to suggest the existence of an *ELSV* type formula, i.e. an intersection theoretic expression that explains the polynomiality properties. This proposal was made in [GJV03] for the specific case of *one-part* double Hurwitz numbers, where there are no wall-crossing issues. After [CJM11], an intriguing strengthening of Goulden-Jackson-Vakil's original conjecture was proposed.

Conjecture (Goulden-Jackson-Vakil+) For $\mathbf{x} \in \mathbb{Z}^n$ with $\sum x_i = 0$,

$$H_g(\mathbf{x}) = \int_{\overline{P}(\mathbf{x})} \frac{1 - \Lambda_2 + \ldots + (-1)^g \Lambda_{2g}}{\prod(1 - x_i\psi_i)}, \tag{7.9}$$

where,

(1) $\overline{P}(\mathbf{x})$ is a moduli space (depending on $\mathbf{x}$) of dimension $4g - 3 + n$.
(2) $\overline{P}(\mathbf{x})$ is constant on each chamber of polynomiality.
(3) The parameter space for double Hurwitz numbers can be identified with a space of stability conditions for a moduli functor and the $\overline{P}(\mathbf{x})$ with the corresponding compactifications.

(4) Λ_{2i} are tautological Chow classes of degree $2i$.

(5) ψ_i's are cotangent line classes.

Goulden, Jackson and Vakil, in the one part double Hurwitz number case, propose that the mystery moduli space may be some compactification of the universal Picard stack over $\overline{\mathcal{M}}_{g,n}$. They verify that such a conjecture holds for genus 0 and for genus 1 by identifying $\overline{Pic}_{1,n}$ with $\overline{\mathcal{M}}_{1,n+1}$.

A lot of progress has happened since in terms of understanding the geometry of compactifications of the Picard stack as well as its tautological intersection theory, see e.g. [MV14]; yet to this day an optimal answer to this conjecture has not been given. Intersection theoretic formulas for double Hurwitz numbers on the moduli spaces of curves have been given [CM14, Lew18] ; however in order to witness piecewise polynomiality some sophisticated geometric inputs such as controlling descendant intersections with double ramification classes [BSSZ15] or Chiodo classes are required.

7.7   Exercises

Exercise 7.1 Compute $H_1^{trop}(d, -d)$, $H_2^{trop}(d, -d)$ and observe that the results are polynomial in d.

Exercise 7.2 Consider the Hurwitz potential for simple Hurwitz numbers (this restriction is not at all important, but it makes the notation a bit less cumbersome), and show how the cut and join obtained by "crashing" a simple branch point into the special one gives rise to a differential operator acting on the Hurwitz potential.

Exercise 7.3 Fill in the details of the proof sketch of the tropical Hurwitz numbers correspondence theorem from Sect. 7.3.

Exercise 7.4 If $f : C_1 \cup C_2 \to T_1 \cup_{n_T} T_2$ is a map of nodal curves, n_T is the node separating T_1 and T_2, and $n_1, \ldots, n_k$ are the nodes of C mapping to n_T with ramification orders $x_1, \ldots, x_k$; then the degeneration formula assigns "n_T" multiplicity:

- $\prod x_i$ if you consider C_1 and C_2 are marked curves (i.e. if you distinguish all k nodes even if some of the ramification orders are the same;
- $\mathfrak{z}(x_1, \ldots, x_k)$ if C_1 and C_2 are not considered marked curves, i.e. if two nodes with the same ramification order are indistinguishable.

Now for the exercise:

1. understand why the two situations are equivalent;
2. understand the multiplicity of the degeneration formula in the second case by interpreting f as a map $T_1 \cup_{n_T} T_2 \to B S_d$ giving rise to a fiber product over the inertia stack of $B S_d$.

Hurwitz Numbers from Piecewise Polynomials

In this lecture we explore how double Hurwitz numbers may be obtained as intersection numbers on moduli spaces of curves. First we review an approach that uses descendant intersections with the double ramification cycle. Next we provide an recent perspective on double Hurwitz numbers, that ties together tropical and logarithmic geometry. The double Hurwitz numbers are obtained as the solution of an intersection problem on a birational modification of the moduli space of curves. The key property of such modification is that the (proper transform of the) double ramification cycle is dimensionally transverse to the boundary of the moduli space, making it possible to find a collection of strata cutting down a zero dimensional cycle of degree equal to $H_g(\mathbf{x})$.

8.1 Double Hurwitz Numbers Through DR

In [CM14], the double Hurwitz number is obtained as the degree of a tautological 0-cycle on $\overline{\mathcal{M}}_{g,n}$. Consider the diagram of spaces:

$$\mathcal{M}(x) = \overline{\widetilde{\mathcal{M}}_g}(x) \xrightarrow{\quad stab \quad} \overline{\mathcal{M}}_{g,n}$$

$$\downarrow br$$

$$\mathcal{M}_{br} = \overline{\mathcal{M}}_{0,2+r}(1,1,\varepsilon,\ldots,\varepsilon)/S_r \qquad (8.1)$$

The double Hurwitz number $H_g(\mathbf{x})$ is the degree of br:

$$H_g(\mathbf{x})[pt.] = br^*([pt])$$

We rewrite this expression in terms of ψ classes. There are three different kinds of ψ classes playing a role in the above diagram:

1. $\hat{\psi}_0$: the psi class on the target space at the relative divisor 0, i.e. the first Chern class of the cotangent line bundle at the relative point 0 on the universal target.
2. $\tilde{\psi}_i$: the psi classes on the space of rubber stable maps at the i-th mark. Remember that we are marking the preimages of the relative divisors.
3. ψ_i is the ordinary psi class on the moduli space of curves.

The following lemmas allow to relate these three types of classes among each other.

Lemma 8.1.1

$$\hat{\psi}^{2g-3+n} = \frac{1}{r!}[pt.]$$

This follows immediately by combining the following facts:

1. Any ψ class on an (ordinary) $\overline{M}_{0,n}$ has top self intersection $1[pt.]$.
2. The fact that the point 0 has weight 1 means that no twigs containing the point 0 get contracted despite the small weigths at the other points. Therefore if we consider the contraction map $c : \overline{M}_{0,r+2} \to \overline{M}_0(1, 1, \varepsilon, \dots, \varepsilon)$, we have that $c^*(\psi_1) = \hat{\psi}_0$.
3. The $r!$ factor comes from the fact that the branch space is a S_r quotient of $\overline{M}_0(1, 1, \varepsilon, \dots, \varepsilon)$.

Lemma 8.1.2

$$br^*(\hat{\psi}_0) = x_i \tilde{\psi}_i$$

where it is understood that the i-th mark is a preimage of 0.

Consider the diagram:

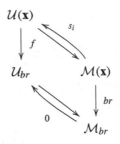

$$(8.2)$$

Then:

$$br^*(\hat{\psi}_0) = -br^*0^*(0) = -s_i^* f^*(0) = -s_i^*(x_i s_i) = x_i \tilde{\psi}_i$$

Combining the two above lemmas, one obtains:

$$H_g(\mathbf{x}) = r! br^*(\hat{\psi}^{2g-3+n}) = r! x_i^{2g-3+n} \tilde{\psi}_i^{2g-3+n}$$

Refer now to Lemma 6.3.3, to see that the tilda-psi classes are pull-backs of ordinary psi classes plus some corrections, namely by the divisor $D_{i,\mathbf{x}}$ in the spaces of relative stable maps parameterizing curves where the mark lies on an unstable component of the curve. Then one can use projection formula to obtain:

$$H_g(\mathbf{x}) = r! x_i^{2g-3+n} (\psi + D_{i,\mathbf{x}})^{2g-3+n} st_*[\overline{\mathcal{M}}_g(\mathbb{P}^1, \mathbf{x})]^{vir} \qquad (8.3)$$

Formula (8.3) explains the piecewise polynomiality of double Hurwitz numbers as follows: intersections of ψ classes with the class $st_*[\widetilde{\mathcal{M}}_g(\mathbb{P}^1, \mathbf{x})]^{vir}$ are shown to be polynomial in the x_i's in [BSSZ15]. The piecewise part arises from the fact that in different chambers of polynomiality one may have different divisorial corrections $D_{i,\mathbf{x}}$.

8.2   Double Hurwitz Numbers as Boundary Intersections

Consider the moduli space $\mathcal{M}_g^{\text{trop},\sim}(\mathbb{P}^1, \mathbf{x})$ of tropical, rubber, relative stable maps, as in [CMR17]; denoting $r = 2g - 2 + n$, there is a branch morphism

$$br_{\text{trop}} : \mathcal{M}_g^{\text{trop},\sim}(\mathbb{P}^1, \mathbf{x}) \rightarrow \left[\overline{M}_{0,2+r}(1, 1, \varepsilon, \ldots, \varepsilon)/S_r\right].$$

Concretely, the target of the branch morphisms may be identified with the parameter space for effective divisors of degree r on $\mathbb{R}$ up to a global translation. Normalizing so that the first point is $0 \in \mathbb{R}$, a fundamental domain corresponds to the cone $\sigma = \{0 \leq t_1 \leq t_2 \leq \ldots \leq t_{r-1}\} \subset \mathbb{R}^{r-1}$.

One has also a stabilization morphism:

$$st_{\text{trop}} : \mathcal{M}_g^{\text{trop},\sim}(\mathbb{P}^1, \mathbf{x}) \to \mathcal{M}_{g,n}^{\text{trop}}.$$

Definition 8.2.1 We denote by $\mathcal{M}_{g,\mathbf{x}}^{\text{trop}}$ the closure of the inverse image via the branch morphism of the interior of the cone σ:

$$\mathcal{M}_{g,\mathbf{x}}^{\text{trop}} := \left(\overline{br_{\text{trop}}^{-1}(\sigma^{\circ})} \right).$$

The cone complex structure on $st_{\text{trop}}(\mathcal{M}_{g,\mathbf{x}}^{\text{trop}})$ is induced from the cone complex structure on $\mathcal{M}_g^{\text{trop},\sim}(\mathbb{P}^1, \mathbf{x})$; notice that in general it does not agree with the one induced by restriction from $\mathcal{M}_{g,n}^{\text{trop}}$. However the integral lattice on $st_{\text{trop}}(\mathcal{M}_{g,\mathbf{x}}^{\text{trop}})$ is restricted from the integral lattice of $\mathcal{M}_{g,n}^{\text{trop}}$.

The space $st_{\text{trop}}(\mathcal{M}_{g,\mathbf{x}}^{\text{trop}})$ is a cone complex of pure codimension g inside $\mathcal{M}_{g,n}^{\text{trop}}$, the closure of the locus parameterizing source curves of tropical covers with no contracting subgraph. The maximal cones of $\mathcal{M}_{g,\mathbf{x}}^{\text{trop}}$ are naturally indexed by monodromy graphs of type $(g, \mathbf{x})$.

The cone complex $st_{\text{trop}}(\mathcal{M}_{g,\mathbf{x}}^{\text{trop}})$ does not necessarily give a subdivision of $\mathcal{M}_{g,n}^{\text{trop}}$, but one may add cones and obtain a subdivision. We denote by $\mathcal{M}_{g,\mathbf{x}}^{\text{trop},+}$ any subdivision of $\mathcal{M}_{g,n}^{\text{trop}}$ which contains $st_{\text{trop}}(\mathcal{M}_{g,\mathbf{x}}^{\text{trop}})$ as a subcomplex. As discussed in Dhruv's mini-course, $\mathcal{M}_{g,\mathbf{x}}^{\text{trop},+}$ determines a birational morphism

$$\pi : \mathcal{M}_{g,\mathbf{x}}^+ \to \overline{\mathcal{M}}_{g,n}, \tag{8.4}$$

and piecewise polynomial functions on $\mathcal{M}_{g,\mathbf{x}}^{\text{trop},+}$ correspond to cohomology classes on $\mathcal{M}_{g,\mathbf{x}}^+$. Also, the proper transform of the closure of the main component of the space of relative stable maps $H_g^{\ell}(\mathbf{x})$ is dimensionally transverse to the boundary of $\mathcal{M}_{g,\mathbf{x}}^+$. For any m-dimensional cone $\sigma \in st_{\text{trop}}(\mathcal{M}_{g,\mathbf{x}}^{\text{trop}})$, there is a corresponding codimension-m, locally closed stratum $\Delta_\sigma \subseteq \mathcal{M}_{g,\mathbf{x}}^+$. The double Hurwitz number $H_g(\mathbf{x})$ is obtained as the intersection of $H_g^{\ell}(\mathbf{x})$ with a union of some of these strata.

Proposition 8.2.2 *We have*

$$H_g(x) = \deg \left(H_g^{\ell}(x) \cdot \sum_{\sigma_\Gamma} \Delta_{\sigma_\Gamma} \right), \tag{8.5}$$

where the sum ranges over all maximal cones of $\mathcal{M}_{g,\mathbf{x}}^{\mathrm{trop}}$, and σ_Γ denotes the cone indexed by the monodromy graph Γ.

Proof   The class of the stratum Δ_{σ_Γ} may be described as a piecewise polynomial function on $\mathcal{M}_{g,\mathbf{x}}^{\mathrm{trop},+}$. Denoting by φ_ρ the piecewise linear function with slope 1 along ρ and zero along all other rays:

$$[\Delta_{\sigma_\Gamma}] = \prod_{\rho \in \sigma_\Gamma} \varphi_\rho =: \varphi_{\sigma_\Gamma}, \tag{8.6}$$

where the product runs over all the rays of the cone σ_Γ.

By projection formula, we have $\deg(H_g^{\ell}(\mathbf{x}) \cdot \Delta_{\sigma_\Gamma}) = \deg(st^*(\varphi_{\sigma_\Gamma}))$. Since the assignment of a cohomology class to a piecewise polynomial function is functorial, we have

$$st^*(\varphi_{\sigma_\Gamma}) = st_{\mathrm{trop}}^*(\varphi_{\sigma_\Gamma}), \tag{8.7}$$

where in the right hand side of (8.7) φ_{σ_Γ} is regarded as a piecewise polynomial function, whereas in the left hand side as a cohomology class on $\mathcal{M}_{g,\mathbf{x}}^+$. By definition of the maps st_{trop} and br_{trop}, one sees that

$$st_{\mathrm{trop}}^*(\varphi_{\sigma_\Gamma})|_{\sigma_\Gamma} = br_{\mathrm{trop}}^* \left(t_1 \prod_{i=2}^{r-1}(t_i - t_{i-1}) \right)_{\Big|_{\sigma_\Gamma}} ; \tag{8.8}$$

whereas for $\tilde{\Gamma} \neq \Gamma$,

$$st_{\mathrm{trop}}^*(\varphi_{\sigma_\Gamma})|_{\sigma_{\tilde{\Gamma}}} = 0. \tag{8.9}$$

Summing over all monodromy graphs, one obtains

$$st_{\mathrm{trop}}^* \left(\sum_\Gamma \varphi_{\sigma_\Gamma} \right) = br_{\mathrm{trop}}^* \left(t_1 \prod_{i=2}^{r-1}(t_i - t_{i-1}) \right). \tag{8.10}$$

Since the polynomial function $t_1 \prod_{i=2}^{r-1}(t_i - t_{i-1})$ on $\left[\overline{M}_{0,2+r}(1, 1, \varepsilon, \ldots, \varepsilon)/S_r \right]$ corresponds to the class of the unique closed stratum in the algebraic moduli space, we have

$$st_{\mathrm{trop}}^* \left(\sum_\Gamma \varphi_{\sigma_\Gamma} \right) = br^* ([pt.]) = H_g(\mathbf{x})[pt.], \tag{8.11}$$

from which the statement of the Proposition follows immediately. $\square$

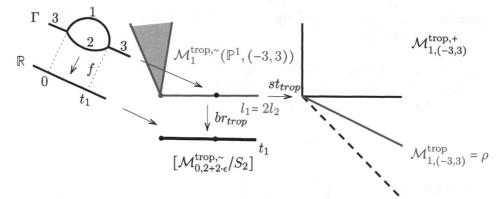

Fig. 8.1 The subdivision of the moduli space $\mathcal{M}_{1,2}^{\text{trop}}$ induced by the moduli space of tropical rubber maps $\mathcal{M}_1^{\text{trop},\sim}(\mathbb{P}^1, (-3,3))$

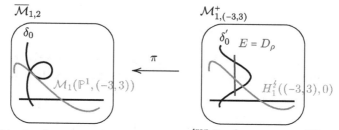

Fig. 8.2 The birational transormation induced by $\mathcal{M}_1^{\text{trop},\sim}(\mathbb{P}^1, (-3,3))$ on $\overline{\mathcal{M}}_{1,2}$ and the dimensionally transverse cycle $H_g^{\frac{t}{2}}(\mathbf{x})$

Example 8.2.3 We consider the double Hurwitz number $H_1((-3,3)) = 2$ and compute it as in Proposition 8.2.2. The moduli space of tropical rubber stable maps $\mathcal{M}_1^{\text{trop},\sim}(\mathbb{P}^1, (-3,3))$, illustrated in red in Fig. 8.1, is not equidimensional: it is the union of a two dimensional cone parameterizing tropical maps with a contracting tropical elliptic curve, with a one dimensional cone, whose general element parameterizes the covers with two compact edges of length l_1 and l_2, as drawn on the leftmost side of Fig. 8.1. The tropical branch morphism contracts the two dimensional cone of $\mathcal{M}_1^{\text{trop},\sim}(\mathbb{P}^1, (-3,3))$ and maps the one dimensional cone onto the unique ray of $[\mathcal{M}_{0,2+2\cdot\epsilon}^{\text{trop},\sim}/S_2]$. The image of the one dimensional cone of $\mathcal{M}_1^{\text{trop},\sim}(\mathbb{P}^1, (-3,3))$ via the tropical stabilization morphism is the the slope $-1/2$ ray ρ in $\mathcal{M}_{1,2}^{\text{trop}}$. This ray alone gives a subdivision $\mathcal{M}_{1,2}^{\text{trop},+}$ of $\mathcal{M}_{1,2}^{\text{trop}}$, which imposes a simple toroidal blowup in the algebraic moduli space $\overline{\mathcal{M}}_{1,2}$, as depicted in Fig. 8.2. The closure of $\mathcal{M}_1^{\text{trop},\sim}(\mathbb{P}^1, (-3,3))$ is not dimensionally transversal to the boundary of $\overline{\mathcal{M}}_{1,2}$, but its proper transform $H_1^{\frac{t}{2}}(-3,3)$ is dimensionally transverse to the exceptional divisor $E = D_\rho$. One may compute the intersection multiplicity $|D_\rho \cdot H_1^{\frac{t}{2}}(-3,3)|$ via a direct local coordinate computation as in Hannah's minicourse (or see [CMR16]). We bypass this technical computation by observing that the piecewise linear function φ_ρ associated to the exceptional divisor D_ρ is such that:

$$st^*_{\text{trop}}(\varphi_\rho) = l_1 = 2l_2 = br^*_{\text{trop}}(t_1). \tag{8.12}$$

The multiplicity of $br^*_{\text{trop}}(t_1)$ equals the product of the weights of the compact edges of the graph Γ.

Alternatively, one may observe that the piecewise linear function t_1 determines the class of a point on the Losev-Manin space $[LM(2)/S_2]$, and the assignment of a cohomology class on the algebraic moduli space to a piecewise polynomial on the tropical moduli space is functorial. It follows from (8.12) that the degree of the class associated to the piecewise polynomial function $st^*_{\text{trop}}(\varphi_\rho)$ equals the degree of $br^*[pt.]$, which is by definition the Hurwitz number $H_1(-3, 3)$.

So far we presented the computation of the double Hurwitz numbers in terms of birational modifications of $\overline{\mathcal{M}}_{g,n}$ induced by tropical geometry which is most natural from a geometric perspective. One may however replace the cycle $H_g^{\notin}(\mathbf{x})$ by $DR_g^{\notin}(\mathbf{x}, 0)$ and obtain the same results.

Theorem 8.2.4 *With notation as throughout this section, we have*

$$H_g(\mathbf{x}) = \deg\left(DR_g^{\notin}(\mathbf{x}, 0) \cdot br^*_{\text{trop}}\left(t_1 \prod_{i=2}^{r-1}(t_i - t_{i-1})\right)\right). \tag{8.13}$$

Proof For any cone σ in the difference

$$\mathcal{M}_g^{\text{trop},\sim}(\mathbb{P}^1, \mathbf{x}) \setminus \mathcal{M}_{g,\mathbf{x}}^{\text{trop}} \tag{8.14}$$

the branch polynomial restricts identically to 0 on σ:

$$br^*_{\text{trop}}\left(t_1 \prod_{i=2}^{r-1}(t_i - t_{i-1})\right)\bigg|_\sigma \equiv 0. \tag{8.15}$$

Then we have:

$$\deg\left(DR_g^{\notin}(\mathbf{x}, 0) \cdot br^*_{\text{trop}}\left(t_1 \prod_{i=2}^{r-1}(t_i - t_{i-1})\right)\right)$$

$$= \deg\left(H_g^{\notin}(\mathbf{x}) \cdot br^*_{\text{trop}}\left(t_1 \prod_{i=2}^{r-1}(t_i - t_{i-1})\right)\right) = H_g(\mathbf{x}), \tag{8.16}$$

with the last equality being the statement of Proposition 8.2.2. $\qquad\qquad\square$

8.2.1   Lower genus Double Hurwitz Numbers and $DR_g^{\ell}(\mathbf{x}, 0)$

The double Hurwitz number $H_g(\mathbf{x})$ is the degree of the class $br^*([pt.])$ in the moduli space of rubber relative stable maps $\mathcal{M}_g^{\text{trop},\sim}(\mathbb{P}^1, \mathbf{x})$. For a general choice of cycle representing the class of a point, the cycle $br^*([pt.])$ is supported on the main component of the moduli spaces of relative stable maps, the closure of the locus parameterizing maps from smooth source curves. We show how we may choose an appropriate subcomplex of the space of tropical relative stable maps that yields cohomology classes on $\mathcal{M}_{g,\mathbf{x}}^+$ whose intersection with $DR_g^{\ell}(\mathbf{x}, 0)$ extracts the double Hurwitz numbers $H_h(\mathbf{x})$, for $h < g$.

Definition 8.2.5 For $n \geq 1$, let T_n denote a graph obtained from a rooted, trivalent tree with $n + 1$ leaves by attaching vertices of genus 1 at every leaf except the root. While it does not matter what trivalent tree one considers, the graph T_n is to be considered fixed.

For every monodromy graph of type $(h, \mathbf{x})$, attach a copy of T_{g-h} on the end labeled by x_1. The graphs so obtained index a cone sub-complex $M_{g,h,\mathbf{x}}^{\text{trop}}$ of the moduli space $\mathcal{M}_g^{\text{trop},\sim}(\mathbb{P}^1, \mathbf{x})$, which we use to define a cohomology class on $\mathcal{M}_{g,\mathbf{x}}^+$ extracing the double Hurwitz number $H_h(\mathbf{x})$.

Proposition 8.2.6 *We have*

$$H_h(\mathbf{x}) = x_1 \cdot 24^{g-h} \cdot |\text{Aut}(T_{g-h})| \cdot \deg\left(DR_g^{\ell}(\mathbf{x}, 0) \cdot \sum_{\sigma_\Gamma} \Delta_{\sigma_\Gamma} \right), \qquad (8.17)$$

where the sum ranges over all maximal cones of the subcomplex.

Proof Each maximal cone σ_Γ in $M_{g,h,\mathbf{x}}^{\text{trop}}$ has dimension $2g - 3 + n$, and hence the corresponding stratum must intersect $DR_g^{\ell}(\mathbf{x}, 0)$ in top degree, as expected. For every cone σ_Γ, the pull-back $st^*(\Delta_{\sigma_\Gamma})$ is supported on the component $C_h \cong \widetilde{\mathcal{M}}_{h,1}(\mathbb{P}^1, \mathbf{x}) \times \overline{\mathcal{M}}_{g-h,1}$ of the moduli space of relative stable maps $\widetilde{\mathcal{M}}_g(\mathbb{P}^1, \mathbf{x})$ parameterizing a rubber relative stable map of genus h with attached a contracting curve of genus $g - h$.

The piecewise polynomial function φ_{σ_Γ} decomposes as a product:

$$\varphi_{\sigma_\Gamma} = l \cdot \varphi_{\sigma_{T_{g-h}}} \cdot \varphi_{\sigma_\Gamma \setminus T_{g-h}}, \qquad (8.18)$$

where l denotes the length of the edge between the point of attachment of the contracted tree and the next vertex to the right.

On the component C_h, the pull-back $st_{\text{trop}}^*(l)$ correspond to the cohomology class $x_1 ev_1^*([pt.])$, fixing a point on the target where the component must contract. The class associated to the piecewise polynomial $st_{\text{trop}}^*(\sum \varphi_{\sigma_\Gamma \setminus T_{g-h}})$ agrees with $br_h^*([pt.])$, for the genus h branch morphism on the left factor of C_h. Finally $st^*_{\text{trop}}(\varphi_{\sigma T_{g-h}})$ gives the intersection of the stratum $\Delta_{T_{g-h}}$ with the virtual class of C_h. It follows from [Pan99] that $[C_h]^{vir} \cdot \Delta_{T_{g-h}} = \lambda^R_{g-h|\Delta_{T_{g-h}}}$, where the superscript R denotes that the λ class is pulled-back from the right factor of C_h. In conclusion, we have

$$DR^4_g(\mathbf{x}, 0) \cdot \sum_{\sigma_\Gamma} \Delta_{\sigma_\Gamma} = st^*_{\text{trop}}\varphi_{\sigma_\Gamma} = \deg\left(br^*_h([pt.]) \boxtimes \lambda^R_{g-h|\Delta_{T_{g-h}}} \right), \qquad (8.19)$$

where the class in parenthesis is a class in $\widetilde{\mathcal{M}_h}(\mathbb{P}^1, \mathbf{x}) \times \overline{\mathcal{M}}_{g-h,1}$. The class pulled back from the left factor has degree $H_h(\mathbf{x})$. From the isomorphism

$$\Delta_{T_{g-h}} \cong \prod_{i=1}^{g-h} \overline{\mathcal{M}}_{1,1}/|\text{Aut}(|T_{g-h}|),$$

the fact that the class λ_{g-h} splits as the product of λ_1's on each of the factors, and that λ_1 has degree $1/24$ on $\overline{\mathcal{M}}_{1,1}$, one may conclude:

$$DR^4_g(\mathbf{x}, 0) \cdot \sum_{\sigma_\Gamma} \Delta_{\sigma_\Gamma} = H_h(\mathbf{x}) \cdot \frac{1}{|\text{Aut}(T_{g-h})|} \frac{1}{24^{g-h}}, \qquad (8.20)$$

from which the statement of the proposition follows.   □

8.3 Leaky Hurwitz Numbers

Tropical Hurwitz numbers admit a natural combinatorial generalization: one may modify the balancing condition to allow for some amount of leaking at vertices. One then gets some new combinatorial objects (leaky covers) that one may view as degenerations (and in a precise sense, tropicalizations) of twisted differentials. Then one may seek for a correspondence theorem between the weighted count of leaky covers and some geometric count on spaces related to twisted differentials. The main obstruction arises from the fact that on the algebraic geometric side we are lacking a branch morphism. However the perspective introduced in the previous section allows us to bypass this obstacle.

Definition 8.3.1 For fixed g and $\mathbf{x} = (x_1, \ldots, x_n)$, a graph Γ is a *leaky graph* if:

(1) Γ is a connected, genus g, directed graph. (Here, we do not allow genus at vertices.)
(2) Γ has n ends which are directed inward, and labeled by the expansion factors $x_1, \ldots, x_n$. If $x_i > 0$, we say it is an *in-end*, otherwise it is an *out-end*.
(3) All inner vertices of Γ are 3-valent.

(4) After reversing the orientation of the out-ends, Γ does not have sinks or sources.[1]
(5) The inner vertices are ordered compatibly with the partial ordering induced by the directions of the edges.
(6) Every bounded edge e of the graph is equipped with an expansion factor $w(e) \in \mathbb{N}$. These satisfy the *leaky condition* at each inner vertex: the sum of all expansion factors of incoming edges equals the sum of the expansion factors of all outgoing edges plus one.

Definition 8.3.2 The **leaky Hurwitz number** $\ell_{g,\mathbf{x}}$ equals the sum over all leaky graphs Γ, where each is counted with the product of the expansion factors of its bounded edges:

$$\ell_{g,\mathbf{x}} = \sum_{\Gamma} \frac{1}{|\mathrm{Aut}(\Gamma)|} \prod_{e} \omega(e).$$

We first interpret the leaky Hurwitz number in terms of intersection theory on some tropical moduli spaces.

Definition 8.3.3 (Leaky Cover) Let $\pi : \Gamma \to \mathbb{P}^1_{\mathrm{trop}}$ be a surjective map of metric graphs. We require that π is piecewise integer affine linear, the slope of π on a flag or edge e is a positive integer called the *expansion factor* $\omega(e) \in \mathbb{N}_{>0}$.

For a vertex $v \in \Gamma$, the *left resp. right degree of π at v* is defined as follows. Let f_l be the flag of $\pi(v)$ pointing to the left and f_r the flag pointing to the right. Add the expansion factors of all flags f adjacent to v that map to f_l resp. f_r:

$$d_v^l = \sum_{f \mapsto f_l} \omega(f), \quad d_v^r = \sum_{f \mapsto f_r} \omega(f). \tag{8.21}$$

The map $\pi : \Gamma \to \mathbb{P}^1_{\mathrm{trop}}$ is called a *leaky cover* if for every $v \in \Gamma$

$$d_v^l - d_v^r = \mathrm{val}(v) - 2 + 2g(v).$$

Remark 8.3.4 (Vertex Set) For a leaky cover, we fix a vertex set of Γ and $\mathbb{P}^1_{\mathrm{trop}}$ that is minimal in the following sense: each vertex of $\mathbb{P}^1_{\mathrm{trop}}$ contains a vertex of Γ in its preimage which is of genus bigger 0 or valence bigger 2.

Example 8.3.5 Figure 8.3 shows an example of a leaky cover with its minimal vertex set.

Definition 8.3.6 (Left and Right Degree) The *left (resp. right) degree* of a leaky cover is the multiset of expansion factors of its ends mapping to $-\infty$ (resp. $+\infty$).

---

[1] We do not consider leaves to be sinks or sources.

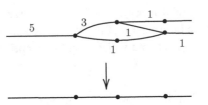

Fig. 8.3 A leaky cover of genus 1 and degree $(5, -1, -1)$, with its minimal vertex set. We do not specify length data in this picture, as the lengths in Γ are imposed by the distances of the points in $\mathbb{P}^1_{\text{trop}}$. For simplicity, we also suppress the labels for the ends in this picture

By convention, we denote the left degree by $\mathbf{x}^+$ and the right degree by $\mathbf{x}^-$. In the right degree, we use negative signs for the expansion factors, in the left degree positive signs. We also merge the two to one vector which we denote $\mathbf{x} = (x_1, \ldots, x_n)$ and call the *degree*. Here, the labeling of the ends plays a role: the expansion factor of the end with the label i is x_i in this notation. In $\mathbf{x}$, we can distinguish the expansion factors of the left ends from those of the right by their sign. It follows from the Euler characteristics of Γ and from the leaky cover condition that

$$\sum_{i=1}^n x_i = n - 2 + 2g,$$

where g denotes the genus of Γ.

An automorphism of a leaky cover is an automorphism of Γ compatible with π.

If we view the expansion factors of a leaky tropical cover as slopes of a rational function on Γ, then the divisor of this rational function is the canonical divisor of Γ.

Definition 8.3.7 (Canonical Divisor) Let Γ be an abstract tropical curve. The *canonical divisor* on Γ is given by $\sum_{v \in \Gamma}(\text{val}(v) - 2 + 2g(v)) \cdot v$.

Definition 8.3.8 (Rational Functions on Abstract Tropical Curves and Their Divisors) Let Γ be an abstract tropical curve. A *rational function* f on Γ is a continuous function $f : \Gamma \to \mathbb{R}$ which is piecewise linear with finitely many pieces and integer slopes. The *order* $\text{ord}_v(f)$ of f at a point v is the sum of the outgoing slopes. The *divisor of a rational function* f is defined as

$$(f) := \sum_{v \in \Gamma} \text{ord}_v(f) \cdot v.$$

Remark 8.3.9 Given a leaky tropical cover $\pi : \Gamma \to \mathbb{P}^1_{\text{trop}}$, we view the expansion factors on the edges as slopes of a rational function f (up to global shift). Then, by definition, the divisor (f) equals the canonical divisor of Γ.

The *combinatorial type* of a leaky cover is the data we obtain when dropping the metric of Γ, i.e. we keep the information of the abstract graph underlying Γ with its genus function, how $\mathbb{P}^1_{\text{trop}}$ is subdivided, and which edge is mapped to which together with the expansion factors.

Remark 8.3.10 The set of all leaky covers of a given combinatorial type forms an open polyhedron in a vector space parametrizing the lengths of all edges. The equations are given by the condition that the cycles have to close up, the inequalities by the fact that edge lengths are positive. We can identify a point on the boundary of such a polyhedron with the cover for which we remove the edges whose lengths have been shrunk to zero. In this way, we can form an *abstract polyhedral complex* parametrizing all leaky covers of genus g and degree $\mathbf{x}$. As common in tropical geometry, the top-dimensional polyhedra in a complex are equipped with a *weight*, which is defined to be the index of the lattice given by the equations that the cycles close up and that vertices have the same image as required times the size of the automorphism group.

Example 8.3.11 Consider the leaky cover from Fig. 8.3 and its combinatorial type. It has four bounded edges, and the equation that the cycle closes up is $3l_1 + l_2 = l_3 + l_4$, where l_1 and l_2 denote the lengths of the upper edges of the cycle and l_3 and l_4 the lengths of the lower. The equation that the two middle vertices have the same image is $3l_1 = l_3$, or, equivalently, $l_2 = l_4$. All leaky covers of this combinatorial type are parametrized by the points in the 2-dimensional open polyhedron

$$\mathbb{R}^4_{>0} \cap \{3l_1 + l_2 - l_3 - l_4 = 0, l_2 - l_4 = 0\}.$$

The index of the lattice defined by $3l_1 + l_2 - l_3 - l_4 = 0, l_2 - l_4 - 0$ equals the greatest common divisor of the absolute values of the 2×2-minors of the matrix

$$\begin{pmatrix} 3 & 1 & -1 & -1 \\ 0 & 1 & 0 & -1 \end{pmatrix}$$

which equals 1. Thus, the weight of the corresponding top-dimensional stratum in the moduli space of leaky covers of genus 1 and degree $(5, -1, -1)$ is 1.

Definition 8.3.12 (Moduli Space of Leaky Covers) We denote the *moduli space of leaky covers of genus g and degree* $\mathbf{x}$, which is the abstract polyhedral complex as described in Remark 8.3.10, by $L_{g,\mathbf{x}}$.

The set of $\mathbb{P}^1_{\text{trop}}$'s subdivided with $r := n - 2 + 2g$ 2-valent genus 0 vertices is an open orthant parametrizing the lengths of the distances between the vertices. We can identify points on the bounday with $\mathbb{P}^1_{\text{trop}}$'s subdivided with fewer vertices. We denote this closed parameter space by P_r.

There is a natural *vertex evaluation map*

$$ev : L_{g,\mathbf{x}} \to \left[\overline{M}_{0,2+r}(1, 1, \varepsilon, \ldots, \varepsilon)/S_r\right],$$

with $r = n - 2 + 2g$.

The degree of a map of weighted abstract polyhedral complexes is the weighted number of preimages of a point in the open interior of P_r. A preimage point is weighted by the product of the weight of the polyhedron in which it lives with the index of the image lattice of this polyhedron under ev in the natural lattice of P_r.

Example 8.3.13 The leaky cover in Fig. 8.3 contributes to the number $N_{1,(5,-1,-1)}$ with the weight 1 times the index of the image lattice of this polyhedron under ev. (The size of its automorphism group is 1.) By Remark 5.19 of [CJM10], this product equals the index of the linear map given by the square matrix in which we combine the equations for the weight with the evaluations. By Example 8.3.11, the polyhedron of the combinatorial type of Fig. 8.3 is surrounded by $\mathbb{R}^4$, thus our matrix is of size 4. Two of its lines is given by the equation of the cycle, $(3, 1, -1, -1)$, and the equation that the middle vertices of Γ have the same image, $(0, 1, 0, -1)$ (see Example 8.3.11). The second vertex of P^1_{trop} is at distance l_3 from the first, the third at distance l_4 from the second. Thus the square matrix to consider is

$$\begin{pmatrix} 3 & 1 & -1 & -1 \\ 0 & 1 & 0 & -1 \\ 0 & 0 & 1 & 0 \\ 0 & 0 & 0 & 1 \end{pmatrix}.$$

The index of the lattice of the image of the linear map defined by this matrix equals the absolute value of its determinant, which is 3. Thus the leaky cover of Fig. 8.3 contributes with weight 3 to the count of $N_{1,(5,-1,-1)}$.

Example 8.3.14 Figure 8.4 shows the count of leaky covers of degree $(5, -1, -1)$ and genus 1. The lowest picture has to be counted twice, as it allows two versions of labeling

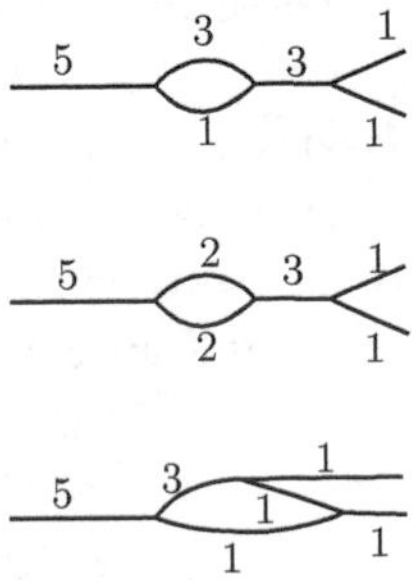

Fig. 8.4 The count of leaky covers of degree $(5, -1, -1)$ and genus 1

its ends. The middle picture has an automorphism group of size 2, as the two edges of the
cycle can be permuted. Thus the total count equalss $9 + 6 + 3 + 3 = 21$.

We conclude by giving a geometric counter-part for the numbers $\ell_{g,\mathbf{x}}$.

First we wish to compactify the open moduli space of meromorphic one forms with a
prescribed divisor. We do so by using logarithmic stable maps.

Informally, let $\mathcal{L}_{g,\mathbf{x}}$ denote the moduli space whose objects are diagrams of the form:

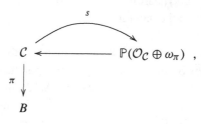

$$(8.22)$$

where $\mathcal{C}$ is a family of semistable n-pointed curves, and s is a section of the projectivization
of the relative dualizing sheaf with orders of contact with the 0 and ∞ section at the marked
points specified by the vector of integers $\mathbf{x}$.

The moduli space $L_{g,\mathbf{x}}$ of tropical leaky covers gives a cone-complex in $\mathcal{M}_{g,n}^{trop}$ (that
we may extend to be a subdivision), and the leaky graphs give naturally a piecewise
polynomial function on $\mathcal{M}_{g,n}^{trop}$. We thus get a cohomology class α on a birational model
of the moduli space of curves. We then obtain the following correspondence theorem.

Theorem 8.3.15 *The intersection between the proper transform of* $st^*(L_{g,\boldsymbol{x}})$ *and* α *has
degree equal to* $\ell_{g,\boldsymbol{x}}$.

8.4   Exercises

Exercise 8.1 The degree of the vertex evaluation map (and thus, the number of leaky
covers of genus g and degree $\mathbf{x}$) is well-defined. All leaky covers in the preimage of a
point in the open interior of P_r are *trivalent covers*, i.e. each preimage of one of the r
vertices of $\mathbb{P}^1_{trop}$ contains precisely one vertex which is not of genus 0 and 2-valent, and
that is of genus 0 and 3-valent.

Exercise 8.2 Let $\pi : \Gamma \to \mathbb{P}^1_{trop}$ be a preimage under the vertex evaluation map of a point
in the open interior of P_r. Then π contributes with the product of the expansion factors
of the bounded edges of Γ times one over the size of the automorphism group of π to the
count of leaky covers.

Exercise 8.3 Conclude from the two previous exercises that the degree of the vertex evaluation map

$$\mathrm{ev}: L_{g,\mathbf{x}} \to \left[\overline{M}_{0,2+r}(1, 1, \varepsilon, \ldots, \varepsilon)/S_r\right],$$

is precisely the leaky Hurwitz number $\ell_{g,\mathbf{x}}$.

Exercise 8.4 Recently many different compactifications of the locus of meromorphic differentials with prescribed divisors have been studied by several groups of people (see [FP18, Sau19]). How do these fit into this story?

Part III

Tropical Plane Curve Counting

Introduction

The MFO-Seminar *Tropical curves, logarithmic structures and enumerative geometry* led by the authors aimed to study the striking relations between algebraic curves, log curves and tropical curves, and their moduli spaces. These relations add to the toolbox of a *enumerative geometry*. In enumerative geometry, we fix a class of geometric objects, mostly curves. We fix some conditions, e.g. a degree of plane curves, or a genus, or points which we require the curves to meet. Then we count how many of our objects actually satisfy the conditions. Of course, we have to choose the conditions in such a way that indeed we obtain a finite-dimensional count and not some higher-dimensional family of objects satisfying the conditions. Often, such enumerative numbers can be expressed in terms of an intersection product on a suitable moduli space: we take a moduli space that parametrizes our objects in question—e.g. the moduli space of plane conics can be described as a $\mathbb{P}^5$ parametrizing the coefficients of a degree 2 homogeneous polynomial in x, y, z. A condition can often be phrased as a subvariety of the moduli space, and the objects satisfying all our conditions are the ones in the intersection of all the subvarieties corresponding to all our conditions. This rough idea motivates why moduli spaces of algebraic curves play a big role in enumerative geometry.

In 2003, following suggestions of Kontsevich, Mikhalkin proved the famous Correspondence Theorem stating the equality of the count $N(d, g)$ of genus g degree d plane curves satisfying point conditions with the analogous count in tropical geometry, $N_{\text{trop}}(d, g)$ [Mik05]. *Tropical geometry* can be viewed as a combinatorial framework to study degenerations in algebraic geometry. The tropicalization of a plane algebraic curve turns out to be a piecewise linear graph in $\mathbb{R}^2$. Following the correspondence theorem, we know that we can count algebraic plane curves by means of enumerating certain piecewise linear graphs in $\mathbb{R}^2$!

One of the main outcomes of Part I is a *correspondence on the level of moduli spaces* [CMR16, CMR17, Ran17]. Logarithmic geometry is a natural link relating moduli spaces which are important for enumerative geometry and their tropical counterparts. The study of tropical moduli spaces, as tools for tropical enumerative geometry, was initiated after Mikhalkin proved the correspondence theorem and has since then led to numerous beautiful results [ACP15, GM08, GKM09, Mik07, CGP21, Uli21, Cap18].

While Part II focuses on Hurwitz theory, i.e. the enumerative geometry of target dimension one, this part is focused on target dimension two. That is, we study counts of plane curves, and we will mostly focus on plane curves that satisfy point conditions. The numbers $N(d, g)$ mentioned above belong to the most prominent enumerative numbers of this course.

The correspondence theorem $N(d, g) = N_{\text{trop}}(d, g)$ can be viewed as the starting point. Having seen that this hands us a way to determine enumerative numbers in algebraic geometry by counting certain graphs in $\mathbb{R}^2$, we ask ourselves how we can exploit this relation to obtain new insights on enumerative geometry. In order to do this, we need to be able to actually count the piecewise linear graphs in question! While this may seem easier than counting algebraic curves, it is still involved, and there are several methods to attack such tropical counting problems.

One of these methods is to use the *duality* of tropical hypersurfaces to subdivisions of a Newton polytope. There is an algorithm to count tropical plane curves by means of this dual world, the so-called *lattice path algorithm* to due Mikhalkin [Mik05]. This algorithm is feasible to count curves in any toric surface. It produces such a count by producing a list of suitable lattice paths, and then computing their multiplicity recursively. Similar to tropical curves, we count lattice paths with a multiplicity, ending up at our desired number. For certain toric surfaces such as $\mathbb{P}^2$, the computation of the multiplicity can be made explicit. This explicit version of the multiplicity then helps to deduce a recursive structure among the lattice paths themselves. The recursion roughly studies the possibilities for the first step of a lattice path, and inserts numbers of lattice paths starting one step later. To make sense of this, we have to extend our notion of lattice paths by allowing paths with different starting points. It turns out that the recursion for lattice paths counting tropical curves of degree d (which, in turn, count plane algebraic curves) equals a famous recursion that was proved by Caporaso and Harris for counts of algebraic plane curves [CH98].

The same recursion can be discovered without referring to lattice paths, but by studying tropical plane curves directly. This has the advantage that we can repeat the recursion, until, ultimately, we end up with a new combinatorial tool to determine counts of plane tropical curves, the so-called *floor diagrams* [BM08, FM10, Blo11]. Floor diagrams are discrete objects which can be counted efficiently.

To sum up, Part III presents several algorithms and methods that are useful to actually produce numbers for certain counts of plane tropical curves. With the correspondence theorem, these tools can be used for the corresponding numbers in algebraic geometry of course. For example, new results on so-called node polynomials (these interpolate counts of plane curves having the same number of nodes) have been achieved via tropical methods [FM10, Blo11]. Also, new results on counts of real algebraic curves have been obtained using the toolbox for tropical plane curve counts presented here [IKS05, ISK09]. We expect the toolbox for tropical plane curve counts to be useful for further interesting research developments in the area.

Introduction to Plane Tropical Curve Counts

9

This lecture is aimed at newcomers to tropical geometry. Topics include plane tropical curves, duality, the relation to algebraic curves via Kapranov's theorem, and the statement of Mikhalkin's correspondence theorem for counts $N_{\text{trop}}(d, g)$ of plane tropical curves of degree d and genus g satisfying point conditions.

Tropical geometry can be viewed as a degenerate version of algebraic geometry (for a precise statement in the case of plane curves, see Kapranov's Theorem 9.3.4). It can also be described as geometry over the max-plus semifield. We introduce it as the latter. Then we deduce important combinatorial features via duality. After that, we come back to the degeneration story. More detailed introductions to tropical geometry can be found e.g. in [BIMS15, BS14, Gat06, IMS09, MS15, RGST05].

After this basic introduction, we focus on counts of tropical plane curves satisfying point conditions. In the end, we present the famous correspondence theorem by Mikhalkin [Mik05], which states that the tropical numbers are equal to their counterparts in algebraic geometry. For example, there is one line in $\mathbb{P}^2$ through 2 points in algebraic geometry, and there is also one tropical line in $\mathbb{R}^2$ through two generic points (see Fig. 9.1, left picture). There is one conic through 5 points in $\mathbb{P}^2$, and also one tropical conic through 5 points in $\mathbb{R}^2$ (see Fig. 9.1, right picture). A first goal is to introduce enough basics such that the pictures of Fig. 9.1 make sense.

9.1 Tropical Polynomials

In algebraic geometry, (complex) plane curves are given as zero-sets of polynomials over the complex numbers. In tropical geometry we can use the analogous definition, by just adding the word "tropical" everywhere: tropical plane curves are tropical vanishing loci of tropical polynomials over the tropical semifield. To make sense of this, we start by defining

Fig. 9.1 A tropical line
through two points and a
tropical conic through five
points in the plane

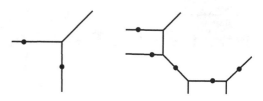

the tropical semifield, tropical polynomials, and then tropical vanishing loci and tropical
plane curves.

Definition 9.1.1 (Tropical Semifield) $(\mathbb{R}\cup\{-\infty\}), \oplus, \odot)$ is called the tropical semifield,
where

$$x \oplus y := \max\{x, y\} \text{ and } x \odot y := x + y.$$

These operations are associative, distributive and commutative. The neutral element for
the addition is $-\infty$. The neutral element for multiplication is 0. Multiplicative inverses are
usual additive inverses, $-x$ for x. But there do not exist additive inverses, as the equation

$$x \oplus a = -\infty \Leftrightarrow \max\{a, x\} = -\infty$$

has no solution x for fixed $a \in \mathbb{R}$. Thus we cannot subtract tropically.

As $(\mathbb{R}\cup\{-\infty\}), \oplus, \odot)$ satisfies all field axioms except the existence of additive inverses,
it is called a semiring or *semifield*.

The tropical semifield is idempotent, i.e. $a \oplus a \oplus \ldots \oplus a = a$. The Freshman's dream
holds tropically:

$$(x \oplus y)^2 = (x \oplus y) \odot (x \oplus y) = \max\{x, y\} + \max\{x, y\} = \max\{2x, 2y\}$$
$$= x \odot x \oplus y \odot y = x^2 \oplus y^2.$$

Remark 9.1.2 Many people use

$$(\mathbb{R} \cup \{\infty\}), \min, +)$$

instead of

$$(\mathbb{R} \cup \{-\infty\}), \max, +).$$

This is isomorphic of course.

Usually, a tropical polynomial is a finite sum of terms of the form

$$a_\alpha \underline{x}^\alpha = a_\alpha \cdot x_1^{\alpha_1} \cdot \ldots \cdot x_n^{\alpha_n}$$

for $\alpha \in \mathbb{N}^n$ and a_α in the ring/field of coefficients. Tropically, we do the same:

Definition 9.1.3 (Tropical Polynomials) A *tropical term* is an expression of the form

$$a_\alpha \odot x_a^{\alpha_1} \odot \ldots \odot x_n^{\alpha_n}$$
$$= a_\alpha \odot (x_1 \odot \ldots \odot x_1) \odot \ldots \odot (x_n \odot \ldots \odot x_n)$$
$$= a_\alpha + (x_1 + \ldots + x_1) + \ldots + (x_n + \ldots + x_n)$$
$$= a_\alpha + \alpha_1 x_1 + \ldots + \alpha_n x_n = a_\alpha + \langle \alpha, x \rangle,$$

where $\langle ., . \rangle$ denotes the Euclidean scalar product on $\mathbb{R}^n$.

Viewed as function $\mathbb{R}^n \to \mathbb{R}$, a tropical term is an affine linear function with rational slope (i.e. $\alpha \in \mathbb{N}^n$).

A *tropical polynomial* is a finite tropical sum of tropical terms, i.e. an expression of the form

$$\max_{\alpha \in \mathbb{N}^n} \{a_\alpha + \alpha_1 x_1 + \ldots + \alpha_n x_n\}.$$

Viewed as function $\mathbb{R}^n \to \mathbb{R}$, a tropical polynomial is a piecewise affine-linear function with finitely many pieces and rational slopes, which is continuous and convex.

Remark 9.1.4 There is a difference between tropical polynomials and tropical polynomial functions.

Example 9.1.5 (Tropical Cubic Univariate Polynomials)
Let

$$f(x) = a \odot x^3 \oplus b \odot x^2 \oplus c \odot x \oplus d.$$

Assume $d - c \le c - b \le b - a$. The graph of such a tropical cubic polynomial is depicted in Fig. 9.2.

By our assumption, all four lines are visible and we have three corner loci, at $x = d - c$, $c - b$, $b - a$. With these, we obtain a factorization of f into linear terms:

$$f = a \cdot (x \oplus (d - c)) \odot (x \oplus (c - b)) \odot (x \oplus (b - a)).$$

The corner loci are therefore also called the *zeros* or *roots* of f.

Fig. 9.2 The graph of a cubic
tropical polynomial

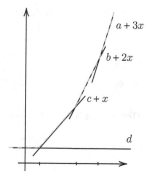

9.2   Tropical Hypersurfaces and Duality

Example 9.1.5 motivates that in general, we view the corner loci of the piecewise linear functions obtained from tropical polynomials as the tropical vanishing loci of tropical polynomials:

Definition 9.2.1 (Tropical Hypersurfaces) Let f be a tropical polynomial in n variables. Then

$$V_{\text{trop}}(f) =$$

$$\{\, x \in \mathbb{R}^n \mid \text{ the maximum of } f \text{ is attained at least by two monomials} \} =$$

$$\text{the corner locus of the piecewise linear function } f \subset \mathbb{R}^n$$

is called the *tropical hypersurface* defined by f. If $n = 2$, we call it a *tropical plane curve*.

Example 9.2.2 Let $f = x \oplus y \oplus 0 = \max\{x, y, 0\}$. Then $V_{\text{trop}}(f)$ is a tropical line, see Fig. 9.3.

Tropical plane curves (and, more generally, tropical hypersurfaces) are *dual* to marked subdivisions of the Newton polytope of f. We introduce such subdivisions and state the theorem.

Fig. 9.3 A tropical line

Definition 9.2.3 Let $Q \subset \mathbb{R}^n$ be a lattice polytope and $\mathcal{A} = Q \cap \mathbb{Z}^n$ the lattice points of Q. Let $\mathcal{A}' \subset \mathcal{A}$. The tuple $(Q, \mathcal{A}')$ is a *marked polytope* if $\mathcal{A}'$ containes the vertices of Q.
A *marked subdivision* of Q is a set $\{(Q_i, \mathcal{A}_i) \mid i = 1, \ldots, k\}$ such that

(1) $(Q_i, \mathcal{A}_i)$ is a marked polytope,
(2) each Q_i is of dimension $\dim(Q)$,
(3) $Q = \bigcup_{i=1}^k Q_i$ is a subdivision of Q, i.e. $Q_i \cap Q_j$ is a (possibly empty) face of both Q_i and Q_j,
(4) $\mathcal{A}_i \subset \mathcal{A}$ for all i,
(5) $\mathcal{A}_i \cap (Q_i \cap Q_j) = \mathcal{A}_j \cap (Q_i \cap Q_j)$.

We call the points in $\mathcal{A}_i$ the marked points, and draw them as black vertices in pictures (the unmarked white).

Using a *height function* (e.g. by defining the coefficients of a tropical polynomial to be the heights) we can define a marked subdivision, the so-called *dual Newton subdivision*, by projecting the upper faces as follows. In order to simplify the notation, we let $n = 2$ in the following. Everything holds analogously for arbitrary n.
    Let

$$p = \bigoplus_\mu c_\mu \odot x^{\mu_1} \odot y^{\mu_2}$$

be a tropical polynomial. Project the upper faces of the *extended Newton polytope*

$$\mathrm{Conv}((\mu_1, \mu_2, c_\mu))$$

to obtain a marked subdivision. Mark a point if it is at the upper boundary of the extended Newton polytope.

Theorem 9.2.4 (Duality) *Let $p(x, y)$ be a tropical polynomial. The tropical curve $V_{\mathrm{trop}}(p)$ associated to p is dual to the Newton subdivision of p in the following sense:*

(1) There is a vertex of $V_{\mathrm{trop}}(p)$ for every polygon in the subdivision.
(2) There is an edge for every edge, the edge in $V_{\mathrm{trop}}(p)$ is perpendicular to its dual edge in the subdivision. Ends of $V_{\mathrm{trop}}(p)$ (unbounded edges) correspond to edges on the boundary.
(3) Each connected component of $\mathbb{R}^2 \setminus V_{\mathrm{trop}}(p)$ corresponds to a vertex of the subdivision.

Using duality, one can draw tropical plane curves quickly, as we demonstrate with the following example.

Fig. 9.4 On the left, the
Newton polygon, with heights,
of the tropical polynomial of
Example 9.2.5. On the right,
the subdivision induced by
these heights via projection
from above

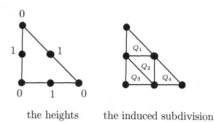

the heights   the induced subdivision

Example 9.2.5 Consider the degree 2 tropical polynomial f given as

$$f = 0 \oplus 1 \odot x \oplus x^2 \oplus 1 \odot y \oplus 1 \odot x \odot y \oplus y^2.$$

The pictures in Fig. 9.4 shows the Newton polygon of this polynomial, i.e. the convex
hull of the exponent vectors. Next to each point, we also note the height which is used to
produce the extended Newton polytope in 3-space. We project the upper faces down to the
Newton polygon. We can see that the points of height 1 form a triangular upper face of
the extended Newton polytope. Two points of height 1 together with a neighbour of height
0 form a triangular face which is bend slightly downwards from this top part of the roof.
Projecting to the Newton polygon, we obtain the picture on the right of Fig. 9.4.

The duality hands us a quick way to draw tropical plane curves. First of all, the
combinatorial type (i.e. the underlying graph and the direction vectors for all edges,
orthogonal to the dual edges) is fixed by the dual Newton subdivision. So, all we need to
know to draw the tropical plane curve, in addition to the subdivision, is the exact position
of the vertices. A vertex is dual to a polygon in the subdivision. The vertex is the locus
where all the terms corresponding to the vertices of this polygon attain their maximum. In
this way, we can determine the positions of the vertices for the example.

The vertices (x, y) dual to the polygons Q_i satisfy:

$$Q_1 : \quad 2y = 1 + y = 1 + x + y \Rightarrow x = 0, y = 1,$$
$$Q_2 : \quad 1 + x = 1 + x + y = 1 + y \Rightarrow x = 0, y = 0,$$
$$Q_3 : \quad 0 = 1 + x = 1 + y \Rightarrow x = -1, y = -1,$$
$$Q_4 : \quad 2x = 1 + x + y = 1 + x \Rightarrow y = 0, x = 1.$$

The tropical plane curve with these vertices is depicted in Fig. 9.5.

It is possible to construct two tropical polynomials that lead to two different subdivi-
sions of the same polygon, but for which the tropical curves coincide as subsets in $\mathbb{R}^2$ (see
Exercise 5). This shows that it is not sufficient to view tropical plane curves as subsets of
$\mathbb{R}^2$. We should equip them with additional structure. First, we add weights on edges. This

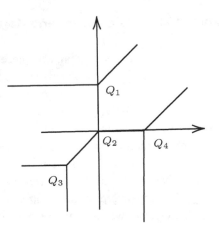

Fig. 9.5 The tropical curve given by the polynomial in Example 9.2.5. It is dual to the subdivision depicted on the right of Fig. 9.4

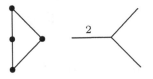

Fig. 9.6 On the right, a tropical plane curve having an edge of weight 2. If we do not mention weights in a picture (as for the remaining two edges here), they are supposed to be one. On the left the dual subdivision having an edge of integer lengths 2 dual to the edge of weight 2

will allow to discover a beautiful feature of tropical plane curves, namely that they satisfy the balancing condition.

Definition 9.2.6 The *weight* of an edge of a tropical plane curve (or, more generally, a tropical hypersurface) equals the lattice length of the corresponding dual edge in the dual Newton subdivision.

For an example, see Fig. 9.6.

Theorem 9.2.7 (Balancing Condition) *Let Γ be a tropical plane curves and $v \in \Gamma$ a vertex. For each edge e adjacent to v, let w_e denote its weight and p_e the primitive integral vector pointing from v into the direction of e. Then*

$$\sum_{v \in \delta e} w_e p_e = 0.$$

Exercise 10 asks for the proof of this theorem.

9.3   Degenerations of Algebraic Curves and Kapranov's Theorem

We can now think of tropical plane curves as weighted piecewise linear graphs in $\mathbb{R}^2$ satisfying the balancing condition. How are these objects related to algebraic plane curves? How can we think about the degeneration taking algebraic curves to tropical curves? Roughly, a degeneration of curves can be viewed as a family of curves over some base with a special fiber. We can just think of a base which is a small disk around 0 such that for all $t \neq 0$, the fibers C_t are isomorphic smooth curves, whereas for $t = 0$ we obtain a degeneration such as e.g. a nodal curve. Each fiber is defined by a (dehomogenized) polynomial in x and y, with t being a parameter. We can view the family as given by a polynomial in t, x and y. We then interpret the parameter t as a coefficient in our underlying field. To make this precise, we work over the field of Puiseux series. This is an example of a field with a non-Archimedean valuation. Other important examples are p-adic numbers or fields of generalized series such as in [Mar10].

Definition 9.3.1 (Puiseux Series) The field $\mathbb{C}\{\{t\}\}$ of *Puiseux series* consists of formal series of the form

$$c(t) = c_1 t^{a_1} + c_2 t^{a_2} + \dots,$$

where the coefficients c_i are in $\mathbb{C}$ and the exponents $a_1 < a_2 < \dots \in \mathbb{Q}$ share a common denominator. (The latter is needed to make the multiplication well-defined.)
     The *valuation map*

$$\mathrm{val} : \mathbb{C}\{\{t\}\} \longrightarrow \mathbb{R} \cup \{\infty\}$$
$$c(t) \longmapsto a_1 \text{ (the least exponent)},$$
$$0 \longmapsto \infty$$

is a non-Archimedean valuation, i.e.

$$\mathrm{val}(x) = \infty \Leftrightarrow x = 0,$$
$$\mathrm{val}(xy) = \mathrm{val}(x) + \mathrm{val}(y),$$
$$\mathrm{val}(x + y) \geq \min\{\mathrm{val}(x), \mathrm{val}(y)\}, \text{ with equality if } \mathrm{val}(x) \neq \mathrm{val}(y).$$

We can view tropicalization as the degeneration taking points over the Puiseux series to their negative valuation.

Definition 9.3.2 (Tropicalization) *Tropicalization* is taking (minus) valuation componentwise:

$$\text{trop} : (\mathbb{C}\{\{t\}\}^*)^2 \longrightarrow \mathbb{R}^2 : (x, y) \mapsto (-val(x), -val(y)).$$

We can also tropicalize a polynomial, by taking minus valuation on the coefficients, and replacing the operations by tropical operations:

Definition 9.3.3 Let $p = \sum c_\mu x^{\mu_1} y^{\mu_2} \in \mathbb{C}\{\{t\}\}[x, y]$. The *tropicalization* of p is defined as

$$\text{trop}(p) = \max\{- \text{val}(c_\mu) + \mu_1 x + \mu_2 y\}.$$

The following theorem tells us that the tropicalization of a plane algebraic curve, i.e. the degeneration of a plane algebraic curve over the Puiseux series, equals a plane tropical curve, which is given by the tropicalization of the defining polynomial. This implies that the tropical plane curves this section starts with, defined via the tropical semiring and tropical polynomials, are not just combinatorial gadgets but receive a meaning as degenerations of algebraic plane curves.

Theorem 9.3.4 (Kapranov's Theorem) *Let $p \in \mathbb{C}\{\{t\}\}[x, y]$. Then*

$$\overline{\text{trop}(V(p))} = V_{\text{trop}}(\text{trop}(p)),$$

i.e. the tropicalization of the algebraic plane curve $V(p)$ (viewed in the torus $(\mathbb{C}\{\{t\}\}^*)^2$) equals the tropical plane curve defined by the tropical polynomial $\text{trop}(p)$. *Here, the bar denotes the closure in the Euclidean topology in $\mathbb{R}^2$.*

Example 9.3.5 Let $p = x + y + t \in \mathbb{C}\{\{t\}\}[x, y]$. Then $V(p)$ is a line and can be parametrized by

$$V(p) = \{(x, -x - t) \mid x \in \mathbb{C}\{\{t\}\}\}.$$

We study the tropicalization of these points.

Case (1): Let $\text{val}(x) > 1$. Then $\text{val}(-x - t) = 1$. The points

$$(- \text{val}(x), - \text{val}(-x - t))$$

of this form yield a ray starting at $(-1, -1) \in \mathbb{R}^2$ and pointing vertically downwards.
Case (2): Let $\text{val}(x) < 1$. Then $\text{val}(-x - t) = \text{val}(x)$. We obtain a ray starting at $(-1, -1)$ and pointing diagonally left upwards.

Case(3): Assume $x = -t + \dots$ Then $-\mathrm{val}(-x - t)$ can be anything bigger than 1. We obtain a ray starting at $(-1, -1)$ and pointing leftwards.

Thus, the tropicalization of the line $V(p)$ is a tropical line with vertex at $(-1, -1)$, and the three rays of direction $(-1, 0)$, $(0, -1)$, and $(1, 1)$.

We can compare with the tropical line given by the tropical polynomial $\max\{-1, x, y\}$ (similar to Example 9.2.2, see Fig. 9.3) and see that the tropicalization of the lines coincides with the tropical line, as predicted by Kapranov's Theorem 9.3.4.

9.4   Parametrizing Tropical Plane Curves

We started by considering plane tropical curves as tropical zero-sets of tropical polynomials. In this way, they obtained the structure of a piecewise linear graph, a subset of $\mathbb{R}^2$. We have seen that it is desirable to add more structure and view tropical plane curves as weighted graphs satisfying the balancing condition. The following example shows that we need even more structure if we want to keep track of the genus of an algebraic curve: we should view tropical curves as being parametrized by an abstract graph.

Example 9.4.1 Consider the cubic polynomial p over the Puiseux series:

$$p = t^5 x^3 + x y^2 + t^3 y^3 - xy - y^2 + (-3t^5) \cdot x + (1 - 3t^3) \cdot y + (2t^3 + 2t^5) \in \mathbb{C}\{\{t\}\}[x, y].$$

Figure 9.7 shows the tropicalization of the cubic (which can be computed using Kapranov's Theorem 9.3.4 and duality as in Example 9.2.5).

The algebraic cubic $V(p)$ has a singularity, a node, at the point $(1, 1)$, as

$$p(1, 1) = \frac{\partial p}{\partial x}(1, 1) = \frac{\partial p}{\partial y}(1, 1) = 0.$$

Fig. 9.7 The tropicalized cubic given by the polynomial in Example 9.4.1 on the right, on the left its dual subdivision

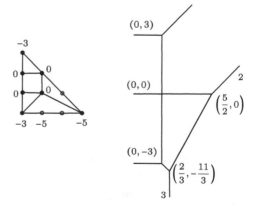

The cubic is thus rational. At first glance, the tropicalization seems to have a cycle however. But we should interpret the vertex dual to the square as a crossing of two edges, not as a vertex. We should parametrize our plane tropical curves by abstract graphs. If we do this, we can use a rational abstract graph, a tree, to parametrize the tropical cubic above.

The example above motivates why we should think of tropical plane curves as being parametrized by an abstract graph, in order to be able to keep track of genus under tropicalization. The abstract graphs are viewed as abtract tropical curves.

In families, cycles of graphs can vanish into vertices (see Remark 9.4.4). The detailed definition of an abstract tropical curve therefore includes a genus function on the vertices keeping track of such issues. For our purposes, it is sufficient to restrict to abstract tropical curves that do not have genus hidden at vertices. Such abstract tropical curves are called explicit.

Definition 9.4.2 (Abstract Tropical Curves) An (explicit) *abstract tropical curve* is a metric graph Γ with unbounded edges called ends of infinite length. The *genus* of a (connected) tropical curve is the first Betti number of the underlying graph.

Due to the Euler characteristics, the genus equals

$$g = 1 - \#\text{vertices} + \#\text{bounded edges}.$$

If Γ is not connected, but consists of connected components $\Gamma_1, \ldots, \Gamma_r$ of genus $g(\Gamma_i) =: g_i$, then the genus of Γ is defined to be

$$g(\Gamma) := \sum_{i=1}^{r} g_i - r + 1.$$

Definition 9.4.3 (Parametrized Tropical Plane Curve/Tropical Stable Map) A *parametrized tropical plane curve* (a.k.a. tropical stable map to $\mathbb{R}^2$) is a tuple (Γ, φ), such that Γ is an abstract tropical curve and φ is a map which is integer affine linear locally on each edge such that the balancing condition is satisfied at every vertex. The weights are given here as expansion factors, i.e. an edge e of length l is mapped to a segment connecting a point $a \in \mathbb{R}^2$ with $a + l \cdot w_e \cdot p_e$, where p_e is the primitive integer vector pointing in the direction of $\varphi(e)$.

The *genus* of (Γ, φ) is defined to be the genus of Γ.

The *degree* of (Γ, φ) is the multiset of directions of its uncontracted ends. We call it *degree d* if the multiset consists of d times the vector $(-1, 0)$, d times $(0, -1)$ and d times $(1, 1)$.

The *combinatorial type* is obtained when forgetting the metric. That is, we keep the topological type of Γ and the weighted directions of all edges.

Fig. 9.8 In a family, "genus can get lost". We can shrink the cycle of the left plane tropical curve to a point, as on the right

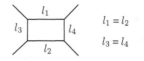

Fig. 9.9 A cycle of a plane tropical curve which imposes two conditions on the lengths of the edges

Remark 9.4.4 (The Moduli Space of Tropical Stable Maps and Its Dimension) The construction of moduli spaces for tropical curves of higher genus poses interesting challenges, since we need to include limits for families of curves for which "genus gets lost" (see Fig. 9.8).

In general, this means we have to allow abstract tropical curves with local genus hidden at vertices. For the purpose of this class, this is not important. We will restrict ourselves to the so-called explicit tropical curves, for which no genus is hidden at vertices.

Let us only get an idea of the dimension of the moduli space of tropical stable maps of genus g and of a fixed degree Δ:

Fix a combinatorial type. If the underlying abstract tropical curve is 3-valent, we have the most bounded edges and thus the most degrees of freedom to vary lengths. One can show by induction that a 3-valent abstract tropical curve of genus g and with $\#\Delta$ ends has $\#\Delta - 3 + 3g$ bounded edges (see Exercise 12).

We cannot vary lengths independently, as cycles have to close up. For an example, see Fig. 9.9.

Every cycle gives two conditions. It is a fact (see e.g. [Mik05]) that for 3-valent plane curves, these $2g$ conditions are independent. Thus, the dimension of the space of tropical stable maps of degree d and genus g is

$$2 + (\#\Delta - 3 + 3g) - 2g,$$

where the first summand accounts for the possibility to globally shift the image $\varphi(\Gamma)$, the second summand equals the number of bounded edges, whose lengths can vary, and the last summand subtracts the conditions we have to satisfy in order for the cycles to close up.

It follows that the dimension of the moduli space of tropical stable maps of degree d and genus g is

$$\#\Delta + g - 1.$$

To construct a well-defined counting problem of tropical stable maps resp. plane tropical curves, we should thus impose $\#\Delta + g - 1$ conditions, e.g. we can fix $\#\Delta + g - 1$ points in $\mathbb{R}^2$ in general position and require our tropical curves to pass through those. If we want to count tropical curves of degree d satisfying point conditions, we thus have to fix $3d + g - 1$ points.

For some piecewise linear graphs in $\mathbb{R}^2$, there are several ways to parametrize them by an abstract tropical curve. Simple tropical curves can uniquely be parametrized, once we fix the convention that every point dual to a parallelogram should be viewed as a crossing of two edges and not as a vertex.

Definition 9.4.5 (Simple Tropical Plane Curves) A parametrized tropical plane curve (Γ, φ) is called *simple* if Γ is 3-valent and the Newton subdivision dual to $\varphi(\Gamma) \subset \mathbb{R}^2$ contains only triangles and paralellograms.

As the following proposition shows, restricting to simple tropical plane curves is ok for the purpose of counting curves satisfying point conditions.

Proposition 9.4.6 *If we fix $3d + g - 1$ generic points in $\mathbb{R}^2$, only simple tropical plane curves of degree d and genus g pass through them.*

The idea of proof is as follows: we equip our parametrized tropical curves with *marked ends*, i.e. we mark some of the unbounded edges and require them to be contracted via φ. Due to the balancing condition, generically a marked end contracts to a point on a line segment in $\varphi(\Gamma)$. We can express the condition to pass through a point $p_i \in \mathbb{R}^2$ as $\varphi(x_i) = p_i$ for a marked end x_i. The points which are met by the marked ends can then be considered as images under the evaluation map sending a tropical stable map to the images of the marked end. The evaluation map can be viewed as a map from the moduli space of tropical stable maps to the space of point conditions. The evaluation map is a piecewise linear map which is linear on the cells corresponding to combinatorial types of the moduli space. If we fix degree d and genus g and $3d + g - 1$ marked ends, then, by Remark 9.4.4, the evaluation map is locally linear of full rank. Thus, lower-dimensional cells of the moduli space (i.e. those corresponding to combinatorial types which are not 3-valent) are mapped to lower-dimensional polyhedra in the space of point conditions. Similarly, we can see that tropical stable maps which are not simple correspond to lower-dimensional parts cut out of top-dimensional cells of the moduli space. The set of *generic point conditions* is formed by the complement of those lower-dimensional polyhedra in the space of point conditions. As its complement is of lower dimension, it is a dense set. By construction, only simple tropical curves pass through generic points.

9.5   The Correspondence Theorem

Tropical curves are counted with *multiplicity*. The multiplicity reflects how many algebraic curves in a counting problem tropicalize to a fixed tropical curve. It is a nice feature of tropical geometry that such a multiplicity can often be computed purely combinatorially. This is the case when we study the count of degree d, genus g plane tropical curves satisfying point conditions. We introduce the multiplicity combinatorially.

Definition 9.5.1 (Multiplicity) Let V be a 3-valent vertex of a plane tropical curve C. Let v, w be two weighted direction vectors of its adjacent edges (see Fig. 9.10). We define the multiplicity of V to be $\text{mult}(V) = |\det(v, w)|$.

By the balancing condition, it does not matter which two of the three adjacent edges we picked.

For a simple tropical curve C we define

$$\text{mult}(C) = \prod_V \text{mult}(V),$$

where the product goes over all 3-valent vertices.

Remark 9.5.2 There is also a global way to express this multiplicity. It can be given as a determinant, which is a tropical intersection multiplicity in the tropical moduli space. The degeneration formula (see Renzo's and Dhruv's lectures) can be viewed as a way to take us from global to local. The local description has the advantage that it is easily computable.

Definition 9.5.3 (The Count of Plane Tropical Curves of Genus g and Degree d Satisfying Point Conditions) Fix $n = 3d + g - 1$ points $p_1, \ldots, p_n \in \mathbb{R}^2$ in general position. Consider parametrized tropical plane curves of degree d and genus g. Let S be the set of such curves (Γ, φ) satisfying $p_i \in \varphi(\Gamma)$ for all i.

Define

$$N_{\text{trop}}(d, g) = \sum_{(\Gamma, \varphi) \in S} \text{mult}(\Gamma, \varphi).$$

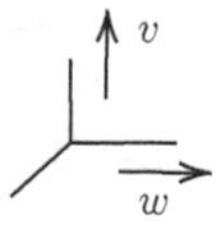

Fig. 9.10 A local picture of a vertex of a tropical curve and the direction vectors of two of its adjacent edges

Here, we do not require the Γ to be connected. If we require that, the number we receive is denoted by $N_{\text{trop}}^{\text{irr}}(d, g)$, where the shortcut irr stands for irreducible.

Theorem 9.5.4 (Mikhalkin's Correspondence Theorem) *Let $N^{\text{irr}}(d, g)$ (resp. $N(d, g)$) denote the numbers of irreducible (resp. not necessarily irreducible) algebraic plane curves of degree d and genus g passing through $3d + g - 1$ points in general position. These numbers equal their tropical counterparts:*

$$N(d, g) = N_{\text{trop}}(d, g)$$

$$N^{\text{irr}}(d, g) = N_{\text{trop}}^{\text{irr}}(d, g).$$

This is a deep theorem. Part I introduces the techniques for the proof in the rational case, involving log moduli spaces.

With the Correspondence Theorem, we can now count algebraic curves by merely counting certain graphs in $\mathbb{R}^2$! This is remarkable. But—how do we count the graphs in $\mathbb{R}^2$, the tropical plane curves, actually? Answers to this question will be given in the next two chapters.

9.6 Exercises

1. Draw the tropical line associated to the polynomial $a \odot x \oplus b \odot y \oplus c$ with $a, b, c \in \mathbb{R} \cup \{-\infty\}$.
2. Show that there is a unique tropical line through two generic points in $\mathbb{R}^2$. What does generic mean here?
3. Draw the tropical curves associated to the tropical polynomials

$$x \odot y \oplus x \oplus y \oplus 0,$$

$$x \odot y \oplus x \oplus y \oplus 1,$$

$$x \odot y \oplus (-1\odot)x \oplus y \oplus 0.$$

4. Compute the dual subdivisions of the tropical curves from Exercise 3 and study their relationship with the curves.
5. Not all subdivisions are dual to plane curves. The one depicted in Fig. 9.11 is not.

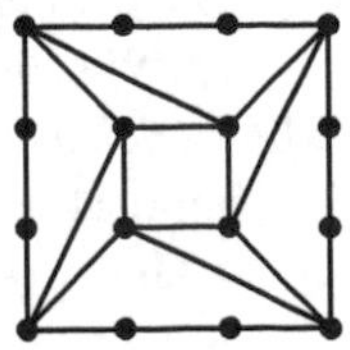

Fig. 9.11 A subdivision which is not dual to a tropical plane curve

Fig. 9.12 Exercise 7 asks for a
tropical polynomial producing
the depicted subdivision

(It might remind one of pictures by the artist Escher you may have heard of. Check
it out if not.)

6. Construct two tropical polynomials producing different subdivisions of the same
   polygon, but for which the associated tropical plane curves (viewed as subsets of $\mathbb{R}^2$)
   coincide. This motivates why we introduce weights for the edges of a tropical plane
   curve.
7. If all the coefficients of a tropical polynomial are equal, what is the dual subdivision?
8. Construct a tropical polynomial that produces the subdivision depicted in Fig. 9.12.
9. Let $p = 0 \oplus a \odot x \oplus x^2$. Describe how the subdivision varies with a.
10. Prove the balancing condition for tropical plane curves (Theorem 9.2.7).
11. Prove (one direction of) Kapranov's Theorem 9.3.4.
12. Show that a 3-valent abstract tropical curve of genus g has $x - 3 + 3g$ bounded edges,
    where x denotes the number of ends.
13. Express the multiplicity of a plane tropical curve in terms of its dual Newton
    subdivision.
14. Construct an example of a simple parametrized tropical plane curve of degree 3 and
    genus 1.
15. Find three parametrized tropical plane curves of degree 3 and of multiplicity 1, 3 resp.
    4. Is there one of multiplicity 2?
16. How can we describe the set of all parametrized tropical plane curves of a fixed
    combinatorial type?
17. What is the genus of a reducible conic? What irreducible components can we have?
    How many reducible conics are there through 4 generic points? Think about this
    question both in algebraic and in tropical geometry.
18. Determine $N_{\text{trop}}(2, 0)$.
19. Let (Γ, φ) be a simple parametrized tropical plane curve. Prove: the genus equals the
    number of interior lattice points that appear in the dual Newton subdivision, minus the
    number of parallelograms that appear in the dual Newton subdivision.
20. Determine $N(2, 0)$ and $N(3, 1)$ (in algebraic geometry). Hint: a plane conic is given
    by a polynomial of the form $ax^2 + bxy + cy^2 + dx + ey + f$ (after dehomogenization).
    The possible coefficients $(a : \ldots : f)$ can be viewed as points in $\mathbb{P}^5$. Fix a point p_1 in
    $\mathbb{P}^2$. Which subset in $\mathbb{P}^5$ parametrizes conics that pass through p_1?
21. The dual of the tropical plane $\mathbb{R}^2$ (i.e. the space parametrizing tropical lines in $\mathbb{R}^2$) can
    be identified with $\mathbb{R}^2$ by sending a tropical line to the negative of its vertex. Observe
    what corresponds to the intersection point of two lines.

22. For fixed g and d, show that there are finitely many combinatorial types of tropical curves in $\mathbb{R}^2$ of this degree and genus.

23. In this lecture, we focused on tropical curves in $\mathbb{R}^2$, i.e. tuples (Γ, φ) with $\varphi : \Gamma \longrightarrow \mathbb{R}^2$, so the dimension of the ambient space is two. Explore the situation in dimension one, i.e. the story of tropical covers. What does the balancing condition say? What is the dimension of the moduli space? Compare with Part II.

Lattice Paths and the Caporaso-Harris Formula   **10**

The lattice path count provides an algorithm to compute the numbers $N_{\text{trop}}(d, g)$ and more general numbers by translating the counting problem to the dual world. The Caporaso-Harris relation, originally found in the algebro-geometric world, naturally appears in the lattice path setting.

The lattice path algorithm to determine the numbers $N_{\text{trop}}(d, g)$ is due to Mikhalkin [Mik05]. Nowadays, the method has been generalized to other counting problems involving tropical plane curves. For example, essentially the same algorithm can be used to determine the famous signed invariant count of real plane curves satisfying real point conditions, the so-called Welschinger-invariant [Mik05, Wel05]. Using this real analogue of the lattice path algorithm, it has been shown that Welschinger invariants are logarithmically equivalent to the analogous count of complex curves [IKS05]. Before, without this tropical approach to counts of real curves, it was not even known whether this invariant is nonzero.

Furthermore, there is a lattice path count to determine counts of real curves passing through real points or pairs of complex conjugate points [Shu06]. There is a lattice path count to determine descendant Gromov-Witten invariants of the plane [MR09], and to determine counts of rational curves satisfying point and cross-ratio conditions [Tyo17, Gol21]. One can use lattice path to determine refined counts of tropical plane curves (these are counts interpolating between real and complex invariants) [GS19, BS19]. The lattice path algorithm has also been generalized to higher dimension, more specifically to counts of tropical surfaces in threespace satisfying point conditions [MMS18]. This (non-exhaustive) list shows the wide range of tropical enumerative problems for which the lattice path method is useful.

This section is built on [GM07, Mar06].

© The Author(s), under exclusive license to Springer Nature Switzerland AG 2023   125
R. Cavalieri et al., *Tropical and Logarithmic Methods in Enumerative Geometry*,
Oberwolfach Seminars 52, https://doi.org/10.1007/978-3-031-39401-0_10

10.1   Lattice Paths

The idea how to translate the enumeration of tropical curves through points in general position to Newton subdivisions is to choose a special (general) position for the points. Then, the marked edges dual to a tropical curve through that special point configuration form a lattice path.

Definition 10.1.1   A path $\gamma : [0, n] \to \mathbb{R}^2$ is called a *lattice path* if $\gamma|_{[j-1,j]}, j = 1, \ldots, n$ is an affine-linear map and $\gamma(j) \in \mathbb{Z}^2$ for all $j = 0 \ldots, n$.

Definition 10.1.2   Let λ be a fixed linear map $\lambda : \mathbb{R}^2 \to \mathbb{R}$ whose kernel has an irrational slope. In the following, we choose $\lambda(x, y) = x - \varepsilon y$, where ε is a small irrational number. A lattice path γ is called λ-*increasing* if $\lambda \circ \gamma$ is strictly increasing.

Let Δ be a Newton polygon of tropical curves. We mostly use triangles of size d, denoted Δ_d, giving us tropical curves of degree d.

Let p and q be the points in Δ where $\lambda|_\Delta$ reaches its minimum (resp. maximum). These points divide the boundary $\partial\Delta$ into two λ-increasing lattice paths $\delta_+ : [0, n_+] \to \partial\Delta$ (going clockwise around $\partial\Delta$) and $\delta_- : [0, n_-] \to \partial\Delta$ (going counterclockwise around $\partial\Delta$), where $n_\pm$ denotes the number of integer points in the two parts of the boundary.

Figure 10.1 shows an example for the triangle Δ_3 with vertices $(0, 0)$, $(3, 0)$ and $(0, 3)$ and the map $\lambda(x, y) = x - \varepsilon y$. The image of the path δ_- is drawn as a line, the image of δ_+ as a dotted line.

Definition 10.1.3   Let $\gamma : [0, n] \to \Delta$ be a λ-increasing path from p to q, that is, $\gamma(0) = p$ and $\gamma(n) = q$. The *(positive and negative) multiplicities* $\mu_+(\gamma)$ and $\mu_-(\gamma)$ are defined recursively as follows:

(1) $\mu_\pm(\delta_\pm) := 1$.
(2) If $\gamma \neq \delta_\pm$ let $k_\pm \in [0, n]$ be the smallest number such that γ makes a left turn (respectively a right turn for μ_-) at $\gamma(k_\pm)$. (If no such $k_\pm$ exists we set $\mu_\pm(\gamma) := 0$). We define two other λ-increasing lattice paths γ' and γ'' as follows:

Fig. 10.1 The triangle Δ_3 and the two paths δ_+ and δ_- on its boundary

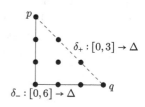

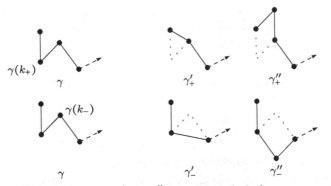

Fig. 10.2 The two λ-increasing paths γ' and γ'' formed recursively from γ

- $\gamma'_{\pm} : [0, n-1] \to \Delta$ is the path that cuts the corner of $\gamma(k_{\pm})$, i.e. $\gamma'_{\pm}(j) := \gamma(j)$ for $j < k_{\pm}$ and $\gamma'_{\pm}(j) := \gamma(j+1)$ for $j \geq k_{\pm}$.
- $\gamma''_{\pm} : [0, n] \to \Delta$ is the path that completes the corner of $\gamma(k_{\pm})$ to a parallelogram, i.e. $\gamma''_{\pm}(j) := \gamma(j)$ for all $j \neq k_{\pm}$ and $\gamma''_{\pm}(k_{\pm}) := \gamma(k_{\pm}-1) + \gamma(k_{\pm}+1) - \gamma(k_{\pm})$.

The two options are depicted in Fig. 10.2.

Let T be the triangle with vertices $\gamma(k_{\pm}-1), \gamma(k_{\pm}), \gamma(k_{\pm}+1)$. Then we set

$$\mu_{\pm}(\gamma) := 2 \cdot \text{Area } T \cdot \mu_{\pm}(\gamma'_{\pm}) + \mu_{\pm}(\gamma''_{\pm}).$$

As both paths $\gamma'_{\pm}$ and $\gamma''_{\pm}$ include a smaller area with $\delta_{\pm}$, we can assume that their multiplicity is known. If $\gamma''_{\pm}$ does not map to Δ, $\mu_{\pm}(\gamma''_{\pm})$ is defined to be zero.

Finally, the *multiplicity* $\mu(\gamma)$ is defined to be the product $\mu(\gamma) := \mu_{+}(\gamma)\mu_{-}(\gamma)$.

Note that the only end path which does not count zero is δ_{+} (respectively δ_{-}), see Fig. 10.1. Paths without a left (resp. right) turn, but not equal to $\delta_{\pm}$, or "faster" paths (taking fewer but bigger steps) $\delta' : [0, n'] \to \Delta$ such that $\delta_{+}([0, n_{+}]) = \delta'([0, n'])$ but $n' < n_{+}$ have multiplicity zero.

Remark 10.1.4 One can interpret the recursion to compute the multiplicity of a path in terms of Newton subdivisions of Δ. Let γ be a λ-increasing path from p to q. First, consider the positive multiplicity. The two paths γ and γ' enclose the triangle T. The two paths γ and γ'' enclose a parallelogram. Take two copies of the Newton polygon Δ and mark the triangle in one of them and the parallelogram in the other one. Continuing like this, one obtains several subdivisions of Δ (above the path γ) in triangles and parallelograms. Analogously, one obtains subdivisions below γ when performing the recursion for μ_{-}. Any such subdivision above γ can be combined with any other subdivision below γ. The set of subdivisions which arise like this is called the set of

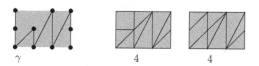

Fig. 10.3 A lattice path γ and its two possible Newton subdivisions. Both are dual to tropical curves of multiplicity 4

possible Newton subdivisions for γ. The multiplicity μ of a path γ is nothing else but the number of possible Newton subdivisions for γ counted with their multiplicity as defined (for dual simple tropical curves) in Chap. 9, Definition 9.5.1. All subdivisions that arise like that contain the edges which are in the image of γ and are simple.

Example 10.1.5 Figure 10.3 shows a λ-increasing path γ (where $\lambda(x, y) = x - \varepsilon y$ as before) and the two possible Newton subdivisions for γ. Both possible subdivisions are dual to tropical curves of multiplicity 4 (because both have two triangles of size 2). The multiplicity of the path is 8.

Definition 10.1.6 Let g be an integer and Δ a convex polygon in $\mathbb{Z}^2$. Let $e = \#(\partial \Delta \cap \mathbb{Z}^2)$. We denote by $N_{\text{path}}(\Delta, g)$ the *number of λ-increasing lattice paths* $\gamma : [0, e + g - 1] \to \Delta$ with $\gamma(0) = p$ and $\gamma(e + g - 1) = q$ counted with their multiplicities, as in Definition 10.1.3.

If $\Delta = \Delta_d$ is the triangle of size d, we also use the notation $N_{\text{path}}(d, g) := N_{\text{path}}(\Delta_d, g)$.

Of course, it is not clear that this definition does not depend on the choice of λ. This follows from Theorem 10.1.7:

Theorem 10.1.7 *Let d be a degree and $g \in \mathbb{Z}$ a genus. As before, we consider lattice paths in the triangle Δ_d of size d. Then*

$$N_{\text{path}}(d, g) = N_{\text{trop}}(d, g),$$

where $N_{\text{path}}(d, g)$ is defined in Definition 10.1.6 and $N_{\text{trop}}(d, g)$ is defined in Definition 9.5.3.

This statement was originally proved in [Mik05], Theorem 2. This theorem can easily be generalized to other polygons, once we generalize the Definition 9.5.3 to count the dual tropical curves appropriately. The idea of proof for this more general version is exactly the same that we present here.

One proves more than the equality of two numbers: one chooses a certain configuration $\mathcal{P}_\lambda$ depending on λ, and then shows that each possible Newton subdivision for a λ-

increasing path (see Remark 10.1.4) is dual to a curve passing through $\mathcal{P}_\lambda$. So in fact, the proof establishes a (weighted) bijection between the tropical curves passing through $\mathcal{P}_\lambda$ and the set of possible Newton subdivisions for all paths.

The following definition concerns the special point configuration needed for the proof of Theorem 10.1.7.

Definition 10.1.8 Choose a map $\lambda : \mathbb{R}^2 \to \mathbb{R}$ as in Definition 10.1.2. Choose a line H orthogonal to the kernel of λ and $n = \#(\partial\Delta \cap \mathbb{Z}^2) + g - 1$ generic points $p_1, \ldots, p_n$ on H such that the distance between p_i and p_{i+1} is much bigger than the distance of p_{i-1} and p_i for all i. Such points $\mathcal{P}_\lambda = \{p_1, \ldots, p_n\}$ are called to be in *Mikhalkin position*.

Let $C = (\Gamma, \varphi)$ be a parametrized tropical plane curve passing through a set of points $\mathcal{P}$. A *string* of C is either a path connecting two ends that does not meet a point of $\mathcal{P}$, or a cycle that does not meet a point of $\mathcal{P}$. If a tropical curve has such a string, we can deform it in an at least one-dimensional family within its combinatorial type without changing the fact that the curves pass through $\mathcal{P}$. For a string which is a path, this can be seen by "pulling" one of its ends a bit to one side, make all its other edges follow according to the balancing condition, and prolong/shorten the edges adjacent to the string accordingly.

Lemma 10.1.9 *Let C be a plane tropical curve through $\mathcal{P}_\lambda$. Then C intersects the line H (on which the points $\mathcal{P}_\lambda = (p_1, \ldots, p_n)$ lie) only in the points $p_1, \ldots, p_n$.*

Proof The general position of $\mathcal{P}_\lambda$ implies that C is simple (see Chap. 9, Proposition 9.4.6) and thus comes from a unique parametrization (Γ, φ). By genericity and the dimension count in Chap. 9, Remark 9.4.4, (Γ, φ) contains no string. Therefore Γ minus the marked points consists of only rational components with one end. If C (and hence $\varphi(\Gamma)$) intersects H also in the point p' different from $p_1, \ldots, p_n$, there must be one component K of Γ minus the marked points whose image intersects H in p', see Fig. 10.4. But then $K \setminus \varphi^{-1}(H)$ consists of two components of which one needs to be compact, as K has only one unbounded end. This is not possible due to the balancing condition. $\qquad\square$

Lemma 10.1.10 *Let C be a tropical curve through points in Mikhalkin position $\mathcal{P}_\lambda$. Let C be of genus g and degree d. Let Ξ denote the edges in the dual Newton subdivision which are dual to the edges passing through $\mathcal{P}_\lambda$.*

Fig. 10.4 Assume a tropical curve met the line H in an additional point p'

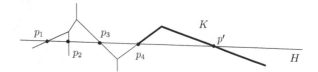

Then Ξ is the image of a λ-increasing path $\gamma : [0, \#(\partial\Delta_d \cap \mathbb{Z}^2) + g - 1] \to \Delta_d$ from p to q (where p and q are, as in Definition 10.1.2, the points of Δ where λ attains its minimum respectively maximum).

Proof First note that by Lemma 10.1.9, the set $\mathcal{P}_\lambda$ coincides with $C \cap H$. Thus one can equivalently show that the edges in the subdivision which are dual to edges of C which intersect H form a lattice path.

Consider a vertex V of the Newton subdivision and the edges adjacent to it. Dual to these edges is a chain of edges of C which encloses a convex polyhedron. Any line meets this chain of edges at most twice. We distinguish several cases depending on the position of V.

(1) Assume first V is in the interior of Δ. Then the convex polyhedron is bounded and H meets either none or two of the dual edges, see Fig. 10.5. It cannot meet a vertex, as $\mathcal{P}_\lambda$ is in general position. Hence, either none or two marked edges must be adjacent to V.

(2) Assume next that $V = p$. Recall how p was defined: it is the vertex of Δ where λ attains its minimum. If we draw a line parallel to the kernel of λ through p then this line meets Δ only in p. Even more, the edges adjacent to p in Δ lie on one side of the line parallel to ker λ. Assume H intersects two edges of C which are adjacent to p. Change the coordinate system for a moment such that H is of slope 0. Then the slope of one edge must be negative and the slope of the other edge must be positive, see Fig. 10.6.

But then the duals of these edges in the Newton polygon Δ would not be on one side of a line parallel to the kernel of λ. So it is not possible that H intersects more than one of the dual edges to the edges adjacent to p.

Assume H intersects none of the dual edges to the interior edges adjacent to p. Then either all those edges lie above H, or below H. Without restriction, assume they lie above H. Then also the ends dual to the edges in the boundary of Δ adjacent to p lie above H. Also, as both of these edges lie on one side of the line parallel to ker λ, one of the dual ends has to intersect H, see Fig. 10.7.

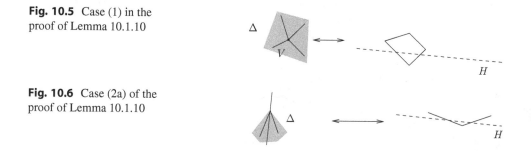

Fig. 10.5 Case (1) in the proof of Lemma 10.1.10

Fig. 10.6 Case (2a) of the proof of Lemma 10.1.10

Fig. 10.7 Case (2b) of the
proof of Lemma 10.1.10

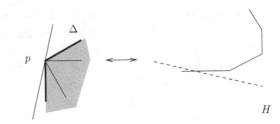

line parallel to ker λ

Hence altogether H intersects precisely one of the duals of the edges adjacent to p.
So there is one marked edge adjacent to p.

(3) Assume that $V = q$, then we get the same result as for p: there is one marked edge
adjacent to q.

(4) Assume finally that V is in the boundary of Δ, but neither equal to p nor q.
    Then the edges adjacent to V do not lie on one side of a line parallel to ker λ through
    V. Therefore we can see as before that H intersects two of these edges, if it intersects
    any at all. So there are either two or no marked edge adjacent to V.

Altogether this shows that at each vertex V—except p and q—there are either two marked
edges adjacent or no marked edge, while at p and q, there is one marked edge adjacent.
That is, Ξ is a path from p to q.

At last, it remains to see that the path Ξ is λ-increasing. Assume the vertices a_1, a_2 and
a_3 are three consecutive vertices of Ξ, such that the step from a_1 to a_2 is λ-increasing,
while the step from a_2 to a_3 is not. But this means, that a_1 and a_3 lie on the same side of a
line parallel to ker λ through a_2. By the above, H cannot intersect both dual edges, which
contradicts the assumption that the two edges were part of Ξ.   □

Proof of Theorem 10.1.7 Take a point configuration $\mathcal{P}_\lambda$ in Mikhalkin position (see
Definition 10.1.8). By definition, it is in general position. By Chap. 9, Remark 9.4.4, the
count of tropical curves passing through these points is finite. Let (Γ, φ) be one of these
curves. By Chap. 9, Proposition 9.4.6, it is simple, due to the general position of $\mathcal{P}_\lambda$. Then
$\varphi(\Gamma)$ is a tropical curve of the right genus and degree through $\mathcal{P}_\lambda$. If we take the edges of
$\varphi(\Gamma)$ that pass through $\mathcal{P}_\lambda$ and consider their dual edges in the Newton subdivision then
these dual edges form a λ-increasing path from p to q by Lemma 10.1.10.

Let γ be a path. The following shows that there are exactly mult(γ) tropical curves
(counted with multiplicity) through $\mathcal{P}_\lambda$, such that the marked points are dual to $\Xi = \gamma$.

Interpret Im(γ) as a set of marked edges in the subdivision and try to draw a dual
tropical curve. For the edges passing through the points of $\mathcal{P}_\lambda$, the direction is prescribed
by the path γ. For each, a point through which it should pass is prescribed by $\mathcal{P}_\lambda$. Take the
first marked edge of $\Xi = $ Im(γ)—that is, the one starting at p—and draw a (part of a) line
orthogonal to this marked edge through p_1. Going on, draw a line dual to the following

marked edge of $\Xi = \mathrm{Im}(\gamma)$ through p_2 and so on. Treat these line segments as the germs of edges of a tropical curve dual to a possible subdivision for Ξ. Recursively, the edge germs grow into a tropical curve dual to a possible subdivision for Ξ. To prove the result, one has to count the possibilities for this. There will be one tropical curve dual to each possible Newton subdivision for γ (see Remark 10.1.4).

The multiplicity μ_+ counts the possibilities to complete the edges dual to γ to a tropical curve (times their multiplicity) in the half-plane above H, whereas μ_- counts below H.

The following makes this argument precise for μ_+, for μ_- it is analogous.

Let the first left turn of the path γ be enclosed by the edges E and E' whose duals e and e' pass through p_i and p_{i+1}. The edges through the points $p_1, \ldots, p_{i-1}$ do not intersect above H, as this was the first left turn. The edges e and e' intersect above H, but below all other possible intersections of dual edges of $\Xi = \mathrm{Im}(\gamma)$. This is true due to the chosen configuration of points: the distance of p_{j+1} and p_j is much bigger than the distance of p_j and p_{j-1}. That is, one can draw a parallel line H' to H such that H and H' enclose a strip in which only the intersection point of e and e' lies. Passing from γ to γ' and γ'' corresponds to moving the line H up to H'. The path γ' leaves a triangle T out, and γ'' completes the corner to a parallelogram. These two possibilities are dual to the two possibilities how a tropical curve can look like at $e \cap e'$: it can either have a 3-valent vertex—in which case it is dual to the triangle T—or e and e' can just intersect—in which case it is dual to the parallelogram which is enclosed by γ and γ''. So the change from γ to γ' and γ'' describes the possibilities how a simple tropical curve through $\mathcal{P}_\lambda$ can look like in the strip enclosed by H and H'.

Figure 10.8 shows a path γ and the two paths γ' and γ'', together with the triangle respectively parallelogram which they enclose with γ. Below, the dual curves in the strip enclosed by H and H' are shown.

Fig. 10.8 The possibilities for a simple tropical plane curve in the strip formed by the line H and a parallel line correspond to the two paths γ' and γ'' in the Recursion to compute the multiplicity of a path, see Definition 10.1.3

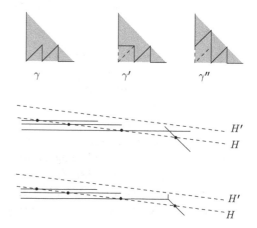

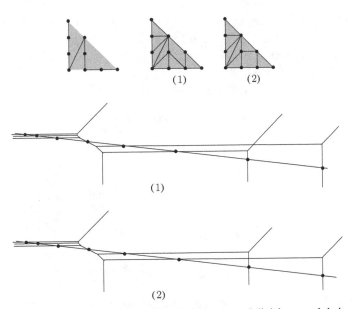

Fig. 10.9 A λ-increasing path, its two possible dual Newton subdivisions, and their corresponding tropical plane curves passing through points in Mikhalkin position

Recursively, we can see that there is in fact exactly one dual tropical curve C through $\mathcal{P}_\lambda$ to each possible Newton subdivision for γ. C is obviously of degree Δ, as we end up with the two paths δ_+ and δ_- which are dual to the ends prescribed by Δ in our recursion.

Hence we constructed a (weighted) bijection between the set of possible Newton subdivisions for a λ-increasing path and the set of tropical curves through $\mathcal{P}_\lambda$. $\square$

Example 10.1.11 Figure 10.9 shows a λ-increasing path $\gamma : [0, 8] \to \Delta_3$ (where $\lambda(x, y) = x - \varepsilon y$ as before). Next to it, its two possible Newton subdivisions are shown, and below the dual curves through the point configuration $\mathcal{P}_\lambda$ in Mikhalkin position. The distance between the points p_{i+1} and p_i is not very much bigger than the distance between p_i and p_{i-1} in the picture, just because this is hard to depict. The picture should still be sufficient to give an idea on how the point configuration in Mikhalkin position and the dual curves for the possible subdivisions for our paths that pass through the points look like. A more accurate picture, for which the distances of the points grow more, can be produced from this by a topological deformation, stretching horizontally.

10.2   The Caporaso-Harris Formula

The Caporaso-Harris formula was originally discovered for counts of algebraic plane curves satisfying point conditions and tangency conditions to a fixed line [CH98].

Notation 10.2.1 *Let $\alpha = (\alpha_1, \alpha_2, \dots)$ be a finite sequence of natural numbers, that is, almost all α_k are zero. If $\alpha_k = 0$ for all $k > n$ we will also write this sequence as $\alpha = (\alpha_1, \dots, \alpha_n)$. For two sequences α and β we define*

$$|\alpha| := \alpha_1 + \alpha_2 + \cdots,$$

$$I\alpha := 1\alpha_1 + 2\alpha_2 + 3\alpha_3 + \cdots,$$

$$I^\alpha := 1^{\alpha_1} \cdot 2^{\alpha_2} \cdot 3^{\alpha_3} \cdot \dots ,$$

$$\alpha + \beta := (\alpha_1 + \beta_1, \alpha_2 + \beta_2, \dots),$$

$$\alpha \geq \beta :\Leftrightarrow \alpha_n \geq \beta_n \text{ for all } n, \text{ and}$$

$$\binom{\alpha}{\beta} := \binom{\alpha_1}{\beta_1} \cdot \binom{\alpha_2}{\beta_2} \cdot \dots .$$

We denote by e_k the sequence which has a 1 at the k-th place and zeros everywhere else.

Definition 10.2.2 *Let $N^{\alpha,\beta}(d, g)$ be a collection of numbers given for each $d \in \mathbb{N}$, $g \in \mathbb{Z}$ and finite sequences α and β with $I\alpha + I\beta = d$. We say that this collection satisfies the Caporaso-Harris formula if*

$$N^{\alpha,\beta}(d, g) = \sum_{k \mid \beta_k > 0} k \cdot N^{\alpha + e_k, \beta - e_k}(d, g)$$

$$+ \sum I^{\beta'-\beta} \cdot \binom{\alpha}{\alpha'} \cdot \binom{\beta'}{\beta} \cdot N^{\alpha',\beta'}(d-1, g')$$

for all d, g, α and β as above with $d > 1$, where the second sum is taken over all α', β' and g' satisfying

$$\alpha' \leq \alpha,$$

$$\beta' \geq \beta,$$

$$I\alpha' + I\beta' = d - 1,$$

$$g - g' = |\beta' - \beta| - 1, \text{ and}$$

$$d - 2 \geq g - g'.$$

10.3   Generalized Lattice Paths

The next goal is to see that the numbers $N_{\text{path}}(d, g)$ satisfy the Caporaso-Harris recursion. To make sense of this, we need a definition of $N_{\text{path}}^{\alpha,\beta}(d, g)$, numbers of *generalized lattice paths*. The main idea is to count the possible Newton subdivisions for a path γ differently from how we have defined them in Remark 10.1.4. One shows that the number of possible Newton subdivisions is equal to the number of *columnwise Newton subdivisions*. These columnwise Newton subdivisions can now be counted column by column—therefore we get a recursive formula by dropping the first column and counting the possibilities in the smaller triangle corresponding to curves of lower degree.

We consider paths in the triangle Δ_d of size d, that do not contain all points of the left boundary. But there can be steps of bigger integer length.

Choose two sequences α and β with $I\alpha + I\beta = d$. Let $\gamma : [0, n] \to \Delta_d$ be a λ-increasing path with $\gamma(0) = (0, d - I\alpha) = (0, I\beta)$ and $\gamma(n) = q = (d, 0)$. The following defines the multiplicity for such a path γ. Again, as in Definition 10.1.3, this multiplicity is the product of a positive and a negative multiplicity defined separately.

Definition 10.3.1 Let $\delta_\beta : [0, |\beta|+d] \to \Delta_d$ be a path such that the image $\delta_\beta([0, |\beta|+d])$ is equal to the piece of boundary of the triangle Δ_d between $(0, I\beta)$ and $q = (d, 0)$, and such that there are β_i steps (i.e. images of a size one interval $[j, j + 1]$) of integer length i at the side $\{x = 0\}$ (and, accordingly, at $\{y = 0\}$ only steps of integer length 1). Define the negative multiplicity $\mu_{\beta,-}(\delta_\beta)$ of all such paths to be 1. Using these end paths the *negative multiplicity* $\mu_{\beta,-}(\gamma)$ of an arbitrary path as above is defined recursively as in Definition 10.1.3 (2).

Figure 10.10 shows all end paths δ_β for $\beta = (2, 1)$ and $d = 5$.

Definition 10.3.2 To compute the *positive multiplicity* $\mu_{\alpha,+}(\gamma)$, extend γ to a path $\gamma_\alpha :$ $[0, |\alpha| + n] \to \Delta_d$ by adding α_i steps of integer length i at $\{x = 0\}$ from $\gamma_\alpha(0) = p$ to $\gamma_\alpha(|\alpha|) = (0, I\beta)$. Then compute $\mu_+(\gamma_\alpha)$ as in Definition 10.1.3 and set $\mu_{\alpha,+}(\gamma) := \frac{1}{I^\alpha} \cdot \mu_+(\gamma_\alpha)$.

Remark 10.3.3 Definition 10.3.2 a priori depends on the order in which we add the α_i steps of lengths i to the path γ to obtain the path γ_α. However, it follows from the alternative description of the positive multiplicity in Proposition 10.3.9 (2) that it is independent.

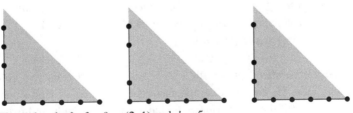

Fig. 10.10 The end paths δ_β for $\beta = (2, 1)$ and $d = 5$

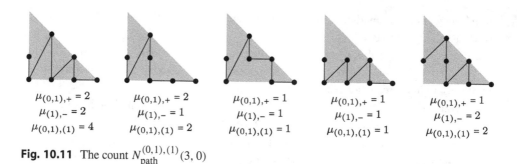

$$\mu_{(0,1),+} = 2 \qquad \mu_{(0,1),+} = 2 \qquad \mu_{(0,1),+} = 1 \qquad \mu_{(0,1),+} = 1 \qquad \mu_{(0,1),+} = 1$$
$$\mu_{(1),-} = 2 \qquad \mu_{(1),-} = 1 \qquad \mu_{(1),-} = 1 \qquad \mu_{(1),-} = 1 \qquad \mu_{(1),-} = 2$$
$$\mu_{(0,1),(1)} = 4 \qquad \mu_{(0,1),(1)} = 2 \qquad \mu_{(0,1),(1)} = 1 \qquad \mu_{(0,1),(1)} = 1 \qquad \mu_{(0,1),(1)} = 2$$

Fig. 10.11 The count $N_{\text{path}}^{(0,1),(1)}(3,0)$

Definition 10.3.4 Let $d \geq 0$ and g be integers, and let α and β be sequences with $I\alpha + I\beta = d$. Define $N_{\text{path}}^{\alpha,\beta}(d,g)$ to be the *number of λ-increasing paths* $\gamma : [0, 2d + g + |\beta| - 1] \to \Delta_d$ that start at $(0, d - I\alpha) = (0, I\beta)$ and end at $(d, 0)$, where each such path is counted with multiplicity $\mu_{\alpha,\beta}(\gamma) := \mu_{\alpha,+}(\gamma) \cdot \mu_{\beta,-}(\gamma)$.

By definition, $N_{\text{path}}(d,g) = N_{\text{path}}^{(0),(d)}(d,g)$. For that reason, if one obtains a recursion for the numbers $N_{\text{path}}^{\alpha,\beta}(d,g)$, this also computes the numbers $N_{\text{path}}(d,g)$ that we cared for from the beginning.

Example 10.3.5 Figure 10.11 shows that

$$N_{\text{path}}^{(0,1),(1)}(3,0) = 4 + 2 + 1 + 1 + 2 = 10.$$

The idea to show that the numbers $N_{\text{path}}^{\alpha,\beta}(d,g)$ satisfy the Caporaso-Harris formula is to determine the possibilities of the first step of each path γ, and to combine with the possibilities how the path can go on. If the first step of the path ends in the second column of Δ_d, one understands the end of the path as a new path in the smaller triangle Δ_{d-1}. For this, one first needs to express the negative and positive multiplicities of a generalized lattice path in a different, non-recursive way. As a preparation, the following lemma characterizes the steps in paths of nonzero multiplicity.

Lemma 10.3.6 *Let α and β be two sequences with $I\alpha + I\beta = d$, and let γ be a generalized lattice path. If γ has a step that moves at least two columns to the right, that is, that starts on a line $\{x = i\}$ and ends on a line $\{x = j\}$ for some i, j with $j - i \geq 2$ then $\mu_{\beta,-}(\gamma) = \mu_{\alpha,+}(\gamma) = \mu_{\alpha,\beta}(\gamma) = 0$.*

Proof Let γ be a generalized path with a step that moves at least two columns to the right. Then both paths $\gamma'_\pm$ and $\gamma''_\pm$ that appear in the recursion of Definition 10.1.3 also contain such a step. Thus the claim follows by induction, as the only end paths of nonzero multiplicity, δ_β (see Definition 10.3.1) and δ_+ (see Definition 10.1.3) of the recursion to compute the multiplicity of γ do not contain such a step. $\square$

Fig. 10.12 Paths with nonzero
multiplicity only contain steps
that go vertically down or
move one column to the right

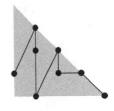

Remark 10.3.7 One concludes that any generalized lattice path with nonzero multiplicity
has two types of steps: some that go down vertically and others that move exactly one
column to the right (with a simultaneous move up or down). An example is depicted in
Fig. 10.12.

Notation 10.3.8 *For a path as in Remark 10.3.7 we fix the following notation: for the
vertical line $\{x = i\}$ in the triangle Δ_d we let $h(i)$ be the highest y-coordinate of a point
of γ on this line. By α^i we denote the sequence describing the lengths of the vertical steps
of γ on this line.*

For example, for the path shown in Fig. 10.12 we have $h(0) = 1$, $h(1) = 3$, $h(2) = 2$,
$h(3) = 1$ and $\alpha^0 = (0)$, $\alpha^1 = (1, 1)$, $\alpha^2 = (1)$, $\alpha^3 = (0)$.

The following interprets both the positive and negative multiplicity of a generalized
lattice path columnwise:

Proposition 10.3.9 *Let α and β be two sequences with $I\alpha + I\beta = d$, and let γ be a
generalized lattice path as in Definition 10.3.4.*

(1) The negative multiplicity of γ is given by the formula

$$\mu_{\beta,-}(\gamma) = \sum_{(\beta^0,\ldots,\beta^d)} \left(\prod_{i=0}^{d-1} I^{\alpha^{i+1}+\beta^{i+1}-\beta^i} \cdot \binom{\alpha^{i+1} + \beta^{i+1}}{\beta^i} \right)$$

*where the sum is taken over all $(d + 1)$-tuples of sequences $(\beta^0, \ldots, \beta^d)$ such that
$\alpha^0 + \beta^0 = \beta$ and $I\alpha^i + I\beta^i = h(i)$ for all i.*

(2) The positive multiplicity of γ is given by the formula

$$\mu_{\alpha,+}(\gamma) = \frac{1}{I^\alpha} \cdot \sum_{(\beta^0,\ldots,\beta^d)} \left(\prod_{i=0}^{d-1} I^{\alpha^i+\beta^i-\beta^{i+1}} \cdot \binom{\alpha^i + \beta^i}{\beta^{i+1}} \right)$$

*where the sum is taken over all $(d + 1)$-tuples of sequences $(\beta^0, \ldots, \beta^d)$ such that
$\beta^0 = \alpha$ and $d - i - I\beta^i = h(i)$ for all i.*

Remark 10.3.10 Before proving Proposition 10.3.9, we interpret its statement geometrically.

The formula of Proposition 10.3.9 (1) counts the number of ways to arrange triangles and parallelograms in Δ_d below γ such that

- the subdivision contains all vertical lines $\{x = i\}$ below γ; and
- each triangle in the subdivision is pointing to the left, that is the vertex opposite to its vertical edge lies to the left of this edge,

where each such subdivision is counted with a multiplicity equal to the product of the double areas of its triangles. We call such a subdivision a *columnwise Newton subdivision*.

The sequences β^i describe the lengths of the vertical edges in the subdivisions below γ. The binomial coefficients $\binom{\alpha^{i+1}+\beta^{i+1}}{\beta^i}$ in the formula count the number of ways to arrange the parallelograms and triangles, and the factors $I^{\alpha^{i+1}+\beta^{i+1}-\beta^i}$ are the double areas of the triangles. As an example let us consider the path of Remark 10.3.7 with $\beta = (1)$. In this case there is only one possibility to fill the area below γ with parallelograms and triangles as above. This is shown in Fig. 10.13, corresponding to $\beta^0 = \beta^2 = \beta^3 = (1), \beta^1 = (0)$. As there is one triangle in this subdivision with double area 2 we see that $\mu_{\beta,-} = 2$.

Note that the original definition of the negative multiplicity of a path (see Definition 10.1.3) also counts Newton subdivisions together with their multiplicity. They were called the *possible Newton subdivisions* for γ in Remark 10.1.4. But the possible Newton subdivisions do not have to be columnwise. In fact, for the path in Fig. 10.13 there is also one possible Newton subdivision of multiplicity 2, but it is not columnwise, as depicted in Fig. 10.14.

The columnwise subdivision of the path γ depicted in Fig. 10.13 does not correspond to a plane tropical curve through Mikhalkin points, whereas the possible subdivision in Fig. 10.14 does. Therefore one can interpret Proposition 10.3.9 as the combinatorial statement that the number of columnwise subdivisions as described above is equal to the

Fig. 10.13 An example of a columnwise subdivision for the region below a path

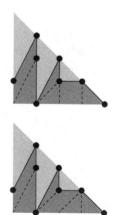

Fig. 10.14 Possible Newton subdivisions for a path do not have to be columnwise

number of possible Newton subdivisions for a path γ (counted with multiplicities), and with that equal to the number of plane tropical curves through points in Mikhalkin position.

For the positive multiplicity there is no such difference between the possible and the columnwise Newton subdivisions: it can be checked that the possible Newton subdivisions for a path γ above γ all contain the vertical lines $\{x = i\}$ above γ and are already columnwise.

Proposition 10.3.9 allows us to split off the first column to obtain a similar subdivision of Δ_{d-1}. This is a key ingredient in the proof of the Caporaso-Harris formula for lattice paths in Theorem 10.3.12.

Proof of Proposition 10.3.9 Start with part (1). The proof is an induction on the recursive computation of $\mu_{\beta,-}$ in Definition 10.3.1. For the induction beginning, consider the end paths in this recursion (the paths that go from $(0, I\beta)$ to $(d, 0)$ along the border of Δ_d). They satisfy the stated formula: all these paths have $\beta^0 = (0)$, so the condition $\alpha^0 + \beta^0 = \beta$ requires $\alpha^0 = \beta$.

Now assume that $\gamma : [0, n] \to \Delta_d$ is an arbitrary generalized lattice path. By induction, the paths γ' and γ'' of Definition 10.1.3 satisfy the formula. Recall that if $k \in [1, n-1]$ is the first vertex at which γ makes a right turn then γ' and γ'' are defined by cutting this vertex $\gamma(k)$ (respectively completing it to a parallelogram). By Lemma 10.3.6, $\gamma(k-1)$ (respectively $\gamma(k+1)$) can be at most one column to the left (respectively right) of $\gamma(k)$. But $\gamma(k-1)$ cannot be in the same column as $\gamma(k)$ as otherwise the λ-increasing path γ could not make a right turn at $\gamma(k)$. Therefore $\gamma(k-1)$ is precisely one column left of $\gamma(k)$. There are two possibilities where $\gamma(k+1)$ can lie:

- $\gamma(k+1)$ can be in the same column i as $\gamma(k)$; or
- $\gamma(k+1)$ can be one column right of $\gamma(k)$.

One proves the statement for these two cases separately.

In the first case, assume that $\gamma(k+1)$ is in column i as $\gamma(k)$. That is, locally the paths γ, γ' and γ'' look as in Fig. 10.15. Then the path γ' has the same values of $h(j)$ and α^j (see Notation 10.3.8) as γ, except for $h(i)$ being replaced by $h(i) - s$ and α^i by $\alpha^i - e_s$, where s is the length of the vertical step from $\gamma(k)$ to $\gamma(k+1)$. The path γ'' has the same

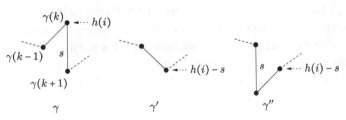

Fig. 10.15 Local pictures of the paths in the recursion

values of $h(j)$ and α^j as γ except for $h(i)$ being replaced by $h(i) - s$, α^i by $\alpha^i - e_s$, and α^{i-1} by $\alpha^{i-1} + e_s$.

Using the formula for γ' and γ'' (which holds by induction), one computes

$$\mu_{\beta,-}(\gamma') = \sum_{(\beta^0,\dots,\beta^d)} \left(\prod_{j=0}^{d-1} I^{\alpha^{j+1}+\beta^{j+1}-\beta^j-\delta_{i,j+1}e_s} \cdot \binom{\alpha^{j+1}+\beta^{j+1}-\delta_{i,j+1}e_s}{\beta^j} \right)$$

$$= \frac{1}{s} \cdot \sum_{(\beta^0,\dots,\beta^d)} \left(\prod_{j=0}^{d-1} I^{\alpha^{j+1}+\beta^{j+1}-\beta^j} \cdot \binom{\alpha^{j+1}+\beta^{j+1}-\delta_{i,j+1}e_s}{\beta^j} \right)$$

(where the sum is taken over the same β as in the statement); and

$$\mu_{\beta,-}(\gamma'') = \sum_{(\beta^0,\dots,\beta^d)} \left(\prod_{j=0}^{d-1} I^{\alpha^{j+1}+\beta^{j+1}-\beta^j} \cdot \binom{\alpha^{j+1}+\beta^{j+1}+\delta_{i-1,j+1}e_s-\delta_{i,j+1}e_s}{\beta^j} \right)$$

where the conditions on the summation variables β^i are $\alpha^0 + \delta_{i-1,0}e_s + \beta^0 = \beta$ and $I(\alpha^j - \delta_{i,j}e_s + \delta_{i-1,j}e_s) + I\beta^j = h(j) - s\delta_{i,j}$. One can make these conditions the same as in the statement by replacing the summation variables β^{i-1} by $\beta^{i-1} - e_s$, arriving at the formula

$$\mu_{\beta,-}(\gamma'') = \sum_{(\beta^0,\dots,\beta^d)} \left(\prod_{j=0}^{d-1} I^{\alpha^{j+1}+\beta^{j+1}-\beta^j} \cdot \binom{\alpha^{j+1}+\beta^{j+1}-\delta_{i,j+1}e_s}{\beta^j-\delta_{i,j+1}e_s} \right).$$

Plugging these expressions into the defining formula

$$\mu_{\beta,-}(\gamma) = s \cdot \mu_{\beta,-}(\gamma') + \mu_{\beta,-}(\gamma'')$$

one arrives at the formula of the statement (where the standard binomial identity $\binom{n-1}{k} + \binom{n-1}{k-1} = \binom{n}{k}$ is used).

For the second case, assume $\gamma(k + 1)$ is one column right of $\gamma(k)$. The idea for this case is the same as for the previous case. But the computation is simpler, because the path γ' does not give a contribution due to Lemma 10.3.6. First note that neither the step from $\gamma(k - 1)$ to $\gamma(k)$ nor the step from $\gamma(k)$ to $\gamma(k + 1)$ can have a negative slope. This is true because otherwise the paths γ'' would also have this negative slope, as the corner is completed to a parallelogram. But the end paths δ_β do not have a step of negative slope, so the claim follows by induction. That is, the heights $h(i)$ of γ in the three columns of $\gamma(k - 1)$, $\gamma(k)$ and $\gamma(k + 1)$ increase.

Fig. 10.16 Local pictures of
paths in the recursion

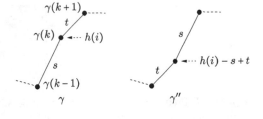

Hence locally at k, the paths γ, γ' and γ'' look as in Fig. 10.16. The path γ'' has the
same values of $h(j)$ and α^j as γ, except for $h(i)$ being replaced by $h(i) - s + t$, where
i is the column of $\gamma(k)$, and s and t are the vertical lengths of the steps before and after
$\gamma(k)$. By the above $s, t \geq 0$. Using the formula of the statement again for γ'', which holds
by induction, one obtains

$$
\mu_{\beta,-}(\gamma'') = \sum_{(\beta^0,\dots,\beta^d)} \left(\prod_{j=0}^{d-1} I^{\alpha^{j+1}+\beta^{j+1}-\beta^j} \cdot \binom{\alpha^{j+1}+\beta^{j+1}}{\beta^j} \right)
$$

where the conditions on the β^j are $\alpha^0 + \beta^0 = \beta$ and $I\alpha^j + I\beta^j = h(j) - (s - t)\delta_{i,j}$.
Note that $s - t > 0$ since γ makes a right turn. As in the previous case one can make the
conditions on the β^j the same as in the statement by replacing the summation variables β^i
by $\beta^{i-1} + \alpha^{i+1} + \beta^{i+1} - 2\beta^i$. Then

$$
\mu_{\beta,-}(\gamma'') =
$$

$$
\sum_{(\beta^0,\dots,\beta^d)} \left(\prod_{j=0}^{d-1} I^{\alpha^{j+1}+\beta^{j+1}-\beta^j} \cdot \binom{\alpha^{j+1}+\beta^{j+1}+\delta_{i,j+1}(\beta^{i-1}+\alpha^{i+1}+\beta^{i+1}-2\beta^i)}{\beta^j + \delta_{i,j}(\beta^{i-1}+\alpha^{i+1}+\beta^{i+1}-2\beta^i)} \right).
$$

This is already the formula of the statement except for the factors

$$
\binom{\alpha^i + \beta^i}{\beta^{i-1}} \binom{\alpha^{i+1} + \beta^{i+1}}{\beta^i}
$$

being replaced by

$$
\binom{\alpha^i + \beta^i + (\beta^{i-1} + \alpha^{i+1} + \beta^{i+1} - 2\beta^i)}{\beta^{i-1}} \binom{\alpha^{i+1} + \beta^{i+1}}{\beta^i + (\beta^{i-1} + \alpha^{i+1} + \beta^{i+1} - 2\beta^i)}
$$

But these terms are the same by the identity $\binom{n+k+l}{n+k}\binom{n+k}{n} = \binom{n+k+l}{n+l}\binom{n+l}{n}$ with $n = \beta^{i-1}$,
$k = \beta^i - \beta^{i-1}$ and $l = \alpha^{i+1} + \beta^{i+1} - \beta^i$, as $\alpha^i = (0)$ for the path γ). This completes the
proof.

Fig. 10.17 Proposition 10.3.9 cannot be generalized to arbitrary polygons. In the picture, the number of columnwise subdivisions would be zero, but the negative multiplicity of the path is one

The proof of case (2) works analogously. However, two steps that move to the next column each can never be the first left turn, because the path starts at p in the first column. This is the reason why the subdivisions above γ are indeed columnwise (see Remark 10.3.10).   □

Remark 10.3.11 It is important for the second step of the proof above that the two boundary lines of Δ_d below and above γ—the line $\{y = 0\}$ respectively the diagonal line from $(0, d)$ to $(d, 0)$—are straight lines. It is not possible to generalize the proof to polygons which contain a vertex above respectively below γ, because then the heights of the three columns of $\gamma(k-1)$, $\gamma(k)$ and $\gamma(k+1)$ cannot be described as $h(i)$, $h(i) + s$ and $h(i) + s + t$. Then the identity $\binom{n+k+l}{n+k}\binom{n+k}{n} = \binom{n+k+l}{n+l}\binom{n+l}{n}$ cannot be used. Figure 10.17 shows a polygon for which the formula of Proposition 10.3.9 does not hold. The columnwise subdivisions would predict 0 as negative multiplicity for the path. However, the path γ'' is a valid end path, so the negative multiplicity is 1. The proof can be generalized to polygons where the boundaries above and below γ are straight lines, for example to rectangles.

Using Proposition 10.3.9, one can now prove the Caporaso-Harris formula for lattice paths:

Theorem 10.3.12 *The numbers $N_{\text{path}}^{\alpha,\beta}(d, g)$ satisfy the Caporaso-Harris formula (see Definition 10.2.2).*

Proof The idea of the proof is to list the possibilities of the first step of the path γ. Let γ be a λ-increasing path from $(0, I\beta)$ to $q = (d, 0)$. As seen in Remark 10.3.7 (using Lemma 10.3.6), there are two cases for the first step of γ (corresponding to the two types of summands in the Caporaso-Harris formula):

- The point $\gamma(1)$ is on the line $\{x = 0\}$.
- The point $\gamma(1)$ is on the line $\{x = 1\}$.

In the first case the point $\gamma(1)$ must be $(0, I\beta - k)$ for some $\beta_k \neq 0$ as otherwise the multiplicity $\mu_{\beta,-}(\gamma)$ would be 0. It follows that the restriction $\gamma|_{[1,2d+g+|\beta|-1]}$ is a path from $(0, d - I(\alpha + e_k))$ and with $\mu_{\alpha,\beta}(\gamma) = k \cdot \mu_{\alpha+e_k,\beta-e_k}(\gamma|_{[1,2d+g+|\beta|-1]})$. Therefore the paths γ with $\gamma(1) \in \{x = 0\}$ contribute

$$\sum_{k:\beta_k>0} k \cdot N_{\text{path}}^{\alpha+e_k,\beta-e_k}(d, g)$$

to the number $N_{\text{path}}^{\alpha,\beta}(d, g)$.

In the second case, use Proposition 10.3.9 to compute both the negative and the positive multiplicity as a product of a factor coming from the first column and the (negative respectively positive) multiplicity of the restricted path $\tilde{\gamma} := \gamma|_{[1,2d+g+|\beta|-1]}$. More precisely, we have

$$\mu_{\alpha,\beta}(\gamma) = \mu_{\beta,-}(\gamma) \cdot \mu_{\alpha,+}(\gamma)$$

$$= \sum_{\beta'} I^{\beta'-\beta}\binom{\beta'}{\beta} \mu_{\beta',-}(\tilde{\gamma}) \cdot \sum_{\alpha'}\binom{\alpha}{\alpha'}\mu_{\alpha',+}(\tilde{\gamma})$$

$$= \sum_{\alpha',\beta'} I^{\beta'-\beta}\binom{\beta'}{\beta}\binom{\alpha}{\alpha'} \cdot \mu_{\alpha',\beta'}(\tilde{\gamma}).$$

So the contribution of the paths with $\gamma(1) \notin \{x = 0\}$ to $N_{\text{path}}^{\alpha,\beta}(d, g)$ is

$$\sum I^{\beta'-\beta}\binom{\beta'}{\beta}\binom{\alpha}{\alpha'} \cdot N_{\text{path}}^{\alpha',\beta'}(d - 1, g')$$

where the sum is taken over all possible α', β' and g'. It remains to determine what these possible values are. It is clear that $\alpha' \le \alpha$ and $\beta \le \beta'$. Also, $I\alpha' + I\beta' = d - 1$ must be fulfilled. As $\tilde{\gamma}$ has one step less than γ, $2d + g + |\beta| - 1 - 1 = 2(d - 1) + g' + |\beta'| - 1$ and hence $g - g' = |\beta' - \beta| - 1$. A path $\epsilon : [0, n] \to \Delta$ from $(0, I\beta)$ to q that meets all lattice points of Δ has $|\beta| + d(d + 1)/2$ steps. As γ has $2d + g - 1 + |\beta|$ steps, $|\beta| + d(d+1)/2 - (2d + g - 1 + |\beta|) = (d - 1)(d - 2)/2 - g$ lattice points are missed by γ. But $\tilde{\gamma}$ cannot miss more points, therefore $(d - 2)(d - 3)/2 - g' \le (d - 1)(d - 2)/2 - g$, that is $d - 2 \ge g - g'$.   □

10.4 Exercises

(1) Count all lattice paths in the size two triangle with 5, and with 4, steps, with multiplicity.
(2) Count all lattice paths in the size three triangle, with 9, and with 8 steps, with multiplicity.
(3) Discuss generalizing the definition of the lattice path count to other polygons.

(4) Count all lattice paths in a size two square with 8 steps, with multiplicity.

(5) Determine $N_{\text{path}}(d, g)$ for $g = \binom{d-1}{2}$ and $g = \binom{d-1}{2} - 1$.

(6) Draw all Newton subdivisions that one obtains by computing the multiplicity of a path contributing to $N_{\text{path}}(3, 0)$, marking the original path. Sketch the dual tropical curves. Is there a configuration of points through which they all fit?

(7) Prove that there exist points in general position (for the counting problem $N_{\text{trop}}(d, g)$) in Mikhalkin position.

(8) Draw 4 points in Mikhalkin position, and draw the tropical conics of genus -1 through them. Compare with Exercise 1. Draw 5 points and all tropical conics of genus 0, compare again.

(9) Draw 8 points and rational cubics, compare with Exercises 2 and 6.

(10) Consider a strip bounded by the line L on which we consider the points in Mikhalkin position, and a parallel shift of L. For the tropical curves you have drawn in the previous exercise, what parts in the dual Newton subdivision correspond to parts in the strip?

(11) Given a tropical curve passing through points in Mikhalkin position, mark the edges in the dual Newton subdivision dual to edges that contain the marked point (contracted ends). Prove that what you obtain is a λ-increasing lattice path.

(12) Discuss again the proof of $N_{\text{trop}}(d, g) = N_{\text{path}}(d, g)$.

(13) How can irreducibility be detected in the lattice path setting?

The Caporaso-Harris Formula for Tropical Plane Curves and Floor Diagrams

11

The Caporaso-Harris recursion has a natural interpretation in terms of tropical plane curves directly. If one applies it ultimately, one obtains a new combinatorial tool to count tropical curves, the so-called floor diagrams. Floor diagrams can be viewed as the combinatorial essence of the tropical curve count.

11.1 The Caporaso-Harris Formula for Tropical Plane Curves

This section is built on [GM07, Mar06]. As in Chap. 10, Sect. 10.3, it is necessary to introduce a more general count of plane curves extending the numbers $N_{\text{trop}}(d, g)$ of Definition 9.5.3 in Chap. 9 to numbers depending also on sequences α and β, as in Chap. 10, Notation 10.2.1.

Definition 11.1.1 Let $d \geq 0$ and g be integers, and let α and β be sequences with $I\alpha + I\beta = d$. Let C be a simple plane tropical curve of genus g and degree

$$\{(\alpha_i + \beta_i) \cdot (-i, 0), d \cdot (0, -1), d \cdot (1, 1) | \ i \in \mathbb{N}\}.$$

That is, C has $\alpha_i + \beta_i$ ends to the left of weight i for all i. Let α_i of these ends have a fixed position, that is, their y-coordinate is fixed.
Define the (α, β)-*multiplicity* of C to be

$$\text{mult}_{\alpha, \beta}(C) := \frac{1}{I^\alpha} \cdot \text{mult}(C)$$

where $\text{mult}(C)$ is the usual multiplicity as in Chap. 9, Definition 9.5.1.

© The Author(s), under exclusive license to Springer Nature Switzerland AG 2023
R. Cavalieri et al., *Tropical and Logarithmic Methods in Enumerative Geometry*,
Oberwolfach Seminars 52, https://doi.org/10.1007/978-3-031-39401-0_11

Fig. 11.1 A tropical curve
having one fixed end, a
(nonfixed) end of weight two,
and passing through six points

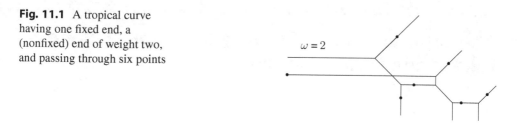

Furthermore, define $N_{\text{trop}}^{\alpha,\beta}(d, g)$ to be the *number of tropical curves* of genus g and degree

$$\{(\alpha_i + \beta_i) \cdot (-i, 0), d \cdot (0, -1), d \cdot (1, 1)| \ i \in \mathbb{N}\}$$

with α_i fixed and β_i non-fixed ends to the left of weight i for all i that pass in addition through a set $\mathcal{P}$ of $2d + g + |\beta| - 1$ points in general position. The curves have to be counted with their respective (α, β)-multiplicities.

Example 11.1.2 Figure 11.1 shows a curve of degree

$$\{(-2, 0), (-1, 0), 3 \cdot (0, -1), 3 \cdot (1, 1)\}.$$

The end of weight 1 to the left is fixed—this is indicated in the picture by a point. There are 6 more points through which the curve passes. As there is one fixed end of weight 1 and one nonfixed of weight 2, the sequences α and β are $\alpha = (1)$ and $\beta = (0, 1)$. The (α, β)-multiplicity of C is equal to the multiplicity of C.

Remark 11.1.3 The ends of higher weight to the left correspond to steps of higher integer length in the boundary of the Newton polygon. Still, one obtains the same Newton polygon as for a curve of degree d. Also, the sum of the directions with primitive integral vector $(-1, 0)$ is $d \cdot (-1, 0)$, because of the assumption $I\alpha + I\beta = d$. Therefore, by abuse of notation those curves are called *of degree d*, even though there are ends of higher weight. This justifies the notation $N_{\text{trop}}^{\alpha,\beta}(d, g)$ for the numbers of curves of degree

$$\{(\alpha_i + \beta_i) \cdot (-i, 0), d \cdot (0, -1), d \cdot (1, 1)| \ i \in \mathbb{N}\}$$

and genus g with the right ends and passing through the prescribed points.

The idea to prove that the numbers $N_{\text{trop}}^{\alpha,\beta}(d, g)$ satisfy the Caporaso-Harris formula is to move one point condition very far left, and to study the possible cases.

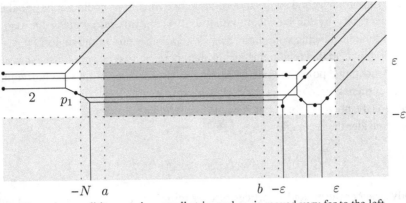

Fig. 11.2 The point conditions are in a small strip, and p_1 is moved very far to the left

Theorem 11.1.4 *The numbers $N_{trop}^{\alpha,\beta}(d, g)$ of tropical plane curves defined in Defini-tion 11.1.1 satisfy the Caporaso-Harris formula (see Definition 10.2.2).*

Proof First, choose a generic point configuration through which the tropical curves are required to pass. Let $\varepsilon > 0$ be a small and $N > 0$ a large real number. Choose the fixed unbounded left ends and the set $\mathcal{P} = \{p_1, \ldots, p_n\}$ so that

- the y-coordinates of all p_i and the fixed ends are in the open interval $(-\varepsilon, \varepsilon)$;
- the x-coordinates of $p_2, \ldots, p_n$ are in the open interval $(-\varepsilon, \varepsilon)$;
- the x-coordinate of p_1 is less than $-N$.

That is, all points are kept in a small horizontal strip and p_1 is moved very far to the left, see Fig. 11.2. Consider a plane tropical curve C satisfying the given conditions. One shows that C must look as in the picture above, that is, that the curve splits into two parts joined by only horizontal lines. Choosing the point configuration in general position implies that C is simple and fixed by the point conditions. Therefore, there is no path from one end to another without meeting a point. Such a path (called a *string*) would lead to a one-dimensional movement of C within its combinatorial type and without losing the property of passing through the points, contradicting the fact that C is fixed by the points. The fact that there is no string is used several times in the following.

First one shows that no vertex of C can have its y-coordinate below $-\varepsilon$. To see this, assume V is a vertex with lowest y-coordinate, and assume it is below $-\varepsilon$. By the balancing condition there must be an edge pointing downwards from V. As there is no vertex below V this must be an end. As the degree is prescribed this end must have direction $(0, -1)$ (especially, weight 1), and it must be the only edge pointing downwards. By the balancing condition it follows that at least one other edge starting at V must be horizontal: If not, both direction vectors of the other two edges adjacent to V would have a nonzero y-coordinate, and so the sum of these two directions could not be equal to $(0, 1)$ which is necessary by the balancing condition. Along this horizontal edge one can go (again due to the balancing condition) from V to another end in the region $\{y \le -\varepsilon\}$. As there are no points in this region this means that there are two ends that we can reach from V without meeting a point (the one pointing downwards from V and the horizontal one). This is a contradiction to the fact that C has no string.

Analogously, no vertex of C can have its y-coordinate above ε.

Next, consider the rectangle

$$R := \{(x, y); \ -N \le x \le -\varepsilon, -\varepsilon \le y \le \varepsilon\}.$$

We study whether there can be vertices of C within R. Let C_0 be a connected component of $C \cap R$. Note that any edge of C_0 leaving R at the top or bottom edge must be unbounded as there are no vertices of C above or below R. If there are edges of C_0 leaving R at the top and at the bottom then one could again go from one end of C to another without passing a point, which is not possible by the above. So without loss of generality C_0 does not meet the top edge of R. With the same argument, C_0 can meet the bottom edge of R only in one point.

Again due to the balancing condition the edges of $C_0 \cap R$ that are not horizontal must then project to the x-axis to a union of two (maybe empty) intervals $[-N, x_1] \cup [x_2, -\varepsilon]$. (Otherwise, there would also be an edge which meets the top edge of R.) But the number of edges of C as well as the minimum slope of an edge (and hence the maximum distance an edge can have within R) are bounded by a constant that depends only on the degree d. So we can find $a, b \in \mathbb{R}, -N \le a < b \le -\varepsilon$ (that depend only on d) such that the interval $[a, b]$ is disjoint from $[-N, x_1] \cup [x_2, -\varepsilon]$, or in other words such that for any curve C which satisfies the conditions, there are no non-horizontal edges in $[a, b] \times [-\varepsilon, \varepsilon]$. Also, there are then no vertices of C in $[a, b] \times \mathbb{R}$. Hence the curve C must look as in Fig. 11.2: one can cut it at any line $x = x_0$ with $a < x_0 < b$ and obtain curves on both sides of this line that are joined only by horizontal edges.

There are now two cases to distinguish, corresponding to the two types of summands which appear in the Caporaso-Harris formula:

(1) p_1 lies on a horizontal non-fixed end of C. Then the region where $x \le -\varepsilon$ consists of only horizontal ends. (If not, there would be at least two ends of direction $(0, -1)$ and $(1, 1)$ in this region, and one would have a path from one to the other without meeting a point, i.e. a string, which is a contradiction.) One can hence consider C as having one more fixed end at p_1 and passing through $\mathcal{P} \setminus \{p_1\}$. One just has to multiply with the weight of this end, as the multiplicity of curves with fixed ends is

defined as $\frac{1}{I^\alpha} \cdot \mathrm{mult}(C)$ by Definition 11.1.1. Therefore the contribution of these curves to $N_{\mathrm{trop}}^{\alpha,\beta}(d, g)$ is

$$\sum_{k:\beta_k>0} k \cdot N_{\mathrm{trop}}^{\alpha+e_k,\beta-e_k}(d, g).$$

This is the first summand in the Caporaso-Harris formula.

(2) p_1 does not lie on a horizontal end of C (as in Fig. 11.2). Then there must also be ends of direction $(0, -1)$ and $(1, 1)$ in this region. Due to the balancing condition we must have as many of direction $(0, -1)$ as of $(1, 1)$. But there is only one point—namely p_1—to separate the ends in this region, therefore there cannot be more than two, as C has no string. So the left part has exactly one end in direction $(0, -1)$ and $(1, 1)$ each, together with some more horizontal ends. It follows that the curve on the right must have degree $d - 1$. Denote this curve by C'.

How many possibilities are there for C? Assume that $\alpha' \leq \alpha$ of the fixed horizontal ends only intersect the part $C \smallsetminus C'$ and are not adjacent to a 3-valent vertex of $C \smallsetminus C'$. Then C' has α' fixed horizontal ends. Given a curve C' of degree $d - 1$ with α' fixed ends through $\mathcal{P} \smallsetminus \{p_1\}$, there are $\binom{\alpha}{\alpha'}$ possibilities to choose which fixed ends of C belong to C'. C' has $d - 1 - I\alpha'$ non-fixed horizontal ends. Let β' be a sequence which fulfills $I\beta' = d - 1 - I\alpha'$, hence a possible choice of weights for the non-fixed ends of C'. Assume that $\beta'' \leq \beta'$ of these ends are adjacent to a 3-valent vertex of $C \smallsetminus C'$ whereas $\beta' - \beta''$ ends intersect $C \smallsetminus C'$, that is, just cross. The connected component of $C \smallsetminus C'$ which contains p_1 has to contain the two ends of direction $(0, -1)$ and $(1, 1)$ due to the balancing condition. (Here connected component means: the image of a connected component of the parametrization.) Also, it can contain some ends of direction $(-1, 0)$—but this have to be fixed ends then, as p_1 cannot separate more than two (nonfixed) ends. So all the β nonfixed ends of direction $(-1, 0)$ have to intersect $C \smallsetminus C'$—therefore they have to be ends of C'. That is, $\beta' - \beta'' = \beta$ (in particular $\beta' \geq \beta$). Given C', there are $\binom{\beta'}{\beta}$ possibilities to choose which ends of C' are also ends of C. Furthermore, one has

$$\mathrm{mult}_{\alpha,\beta}(C) = \frac{1}{I^\alpha} \mathrm{mult}(C) = \frac{1}{I^\alpha} \cdot I^{\alpha-\alpha'} \cdot I^{\beta'-\beta} \cdot \mathrm{mult}(C')$$

$$= \frac{1}{I^{\alpha'}} \cdot I^{\beta'-\beta} \cdot \mathrm{mult}(C') = I^{\beta'-\beta} \cdot \mathrm{mult}_{\alpha',\beta'}(C')$$

where the factors $I^{\alpha-\alpha'}$ and $I^{\beta'-\beta}$ arise due to the 3-valent vertices which are not part of C'. To determine the genus g' of C', note that C' has $|\alpha+\beta''|$ less vertices than C and $|\alpha+\beta''|-1+|\beta''|$ less bounded edges—there are $|\alpha+\beta''|-1$ bounded edges in $C \smallsetminus C'$ and $|\beta''|$ bounded edges are cut. Hence $g' = 1-\#\Gamma^0+|\alpha+\beta''|-\#\Gamma_0^1-|\alpha+\beta''|-|\beta''| = g - (|\beta''| - 1)$. Furthermore, $g - g' \leq d - 2$ as at most $d - 2$ cycles may be cut. Now given a curve C' with α' fixed and β' nonfixed bounded edges through $\mathcal{P} \smallsetminus \{p_1\}$, and having chosen which of the α fixed ends of C are also fixed ends of C' and which of the β' ends of C' are also ends of C, there is only one possibility to add a connected component through p_1 to make it a possible curve C with α fixed ends and β nonfixed. The positions and directions of all bounded edges are prescribed by the point p_1, by the positions of the $\beta' - \beta$ ends to the left of C' and by the $\alpha - \alpha'$ fixed ends. Hence one can count the possibilities for C' (times the factor $\binom{\alpha}{\alpha'} \cdot \binom{\beta'}{\beta} \cdot I^{\beta'-\beta}$) instead of the possibilities for C (where the possible choices for α', β' and g' have to satisfy just the conditions we know from the Caporaso-Harris formula).

The sum of these two contributions gives the required Caporaso-Harris formula.   □

11.2   Floor Diagrams

Applying the Caporaso-Harris algorithm again and again, ultimately we obtain tropical curves which we call floor-decomposed. That is, each connected component of a tropical curve minus its horizontal edges looks like the left part $C \smallsetminus C'$ in Fig. 11.2. The combinatorics of such tropical curves can be condensed even further and we can count them using the so-called *floor diagrams* [BM08, FM10].

Nowadays, the method of floor diagrams is used for numerous tropical enumerative problems. Among them are counts of rational tropical curves satisfying point and cross-ratio conditions [Gol21], Welschinger invariants [ABLdM11], refined curve counts [BG16], Gromov-Witten invariants involving λ-classes [Bou21], stationary descendant Gromov-Witten invariants of Hirzebruch surfaces [CJMR21] and counts of (complex and real) curves in Del Pezzo surfaces [Bru15]. Furthermore, using techniques involving floor diagrams, new results about enumerative problems have been achieved. Among them are new results about node polynomials [FM10, Blo11], structural results about log Gromov-Witten invariants of Hirzebruch surfaces [AB17] and about tropical refined invariants [JPB20], and a tropical approach towards Nagata's conjecture [Kal17]. This (non-exhaustive) list of successes of floor diagrams demonstrates the importance of this tropical counting tool for modern research in enumerative geometry.

Put the point conditions in a small horizontal strip, stretched far apart. Such a point configuration is called in *horizontally stretched* position. (Observe the difference to Mikhalkin position.)

Every plane tropical curve passing through such points is *floor decomposed*, i.e. its dual Newton subdivision contains all vertical line segments. This follows from Theorem 11.1.4 applied recursively.

Definition 11.2.1 (Floors and Elevators) A horizontal edge of a floor-decomposed tropical curve is called an *elevator*, a connected component of the tropical curve minus its elevators is called a *floor*.

Note that one can view points on elevators as floors of size zero. This is consistent with the point of view of having contracted ends for our tropical stable maps which are supposed to meet the point conditions. In this sense, a point on an elevator corresponds to a connected component of the underlying abstract tropical curve which consists of just one contracted end. One can thus consider it a floor of size zero. In the following, to make the text more intuitively accessible, we nevertheless speak of points on elevators rather than of floors of size zero.

The terminology involving floors and elevators is explained by the fact that in parts of the literature, the points are vertically stretched rather than horizontally stretched.

Lemma 11.2.2 *Given a tropical plane curve passing through horizontally stretched points, each floor and each elevator contain precisely one of the points.*

Proof As the point conditions are horizontally stretched, each floor contains at most one point. This aligns with the proof of Theorem 11.1.4 showing how a tropical curve splits when we move one point very far left. If a floor came without a point, there would be a string: a path connecting a vertical end to a diagonal end without meeting a point. One could move this string left or right without changing the fact that the curve meets all the points. The curve would thus not be fixed by the point conditions, which is a contradiction due to the dimension count of the moduli space of tropical stable maps and the fact that the points are chosen in general position. Similarly, an elevator without a marked point would lead to a string. As the points are in general position, there cannot be more than one point on an elevator.   □

The idea is that one obtains the combinatorial essence of the tropical counting problem by shrinking all floors to vertices, and considering the graphs we obtain in this way. Those graphs are called *floor diagrams* and satisfy the properties of the following definition. Figure 11.3 shows a floor decomposed plane tropical curve of degree 5 and genus 0 and its associated floor diagram below.

Definition 11.2.3 (Floor Diagram) A *floor diagram* is a graph on a linearly ordered vertex set, with vertices colored black and white and edges equipped with weights such that

- there are d white vertices and $2d + g - 1$ black vertices,
- d unbounded edges of weight 1 each point to the left,
- each white vertex is of *divergence* 1 (i.e. the sum of the weights of the incoming edges (from the left) equals the sum of the weights of the outgoing edges (to the right) plus 1),
- each black vertex is of valance 2 and divergence 0,
- the graph is bipartite.

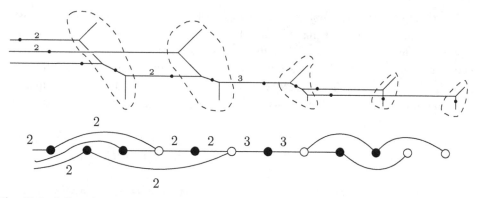

Fig. 11.3 A floor decomposed plane tropical curve of degree 5 and genus 0 and its associated floor diagram

Definition 11.2.4 (Count of Floor Diagrams) Define $N_{\text{floor}}(d, g)$ to be the weighted count of floor diagrams of genus g, with d weights 1 left ends, on $3d + g - 1$ vertices, each counted with multiplicity equal to the product of its edge weights. If one requires the floor diagrams to be connected, the number is denoted by $N_{\text{floor}}^{\text{irr}}(d, g)$.

Theorem 11.2.5 (Floor Diagrams Count Tropical Curves) *Let $d, g \in \mathbb{N}$. Then the numbers of floor diagrams and of plane tropical curves of genus g and degree d satisfying point conditions coincide:*

$$N_{\text{floor}}(d, g) = N_{\text{trop}}(d, g) \text{ and } N_{\text{floor}}^{\text{irr}}(d, g) = N_{\text{trop}}^{\text{irr}}(d, g).$$

Proof The proof of this theorem is in two parts. Construction 11.2.6 associates a floor diagram to a floor decomposed tropical curve. By Proposition 11.2.7, the weighted number of tropical curves yielding the same floor diagram D under this procedure equals the multiplicity mult(D) from Definition 11.2.4. Irreducibility is preserved.   □

Construction 11.2.6 *Let (Γ, φ) be a simple floor decomposed tropical curve of degree d and genus g passing through points in the horizontal strip. Shrink all floors to vertices. Floors of size 0 (points on elevators) are drawn in black, the other floors yield white vertices. The outcome is a floor diagram of degree d and genus g.*

Proof We show that Construction 11.2.6 yields a floor diagram D of the right degree and genus. Because of the horizontally stretched point conditions, we obtain a graph D on a linearly ordered vertex set. The left ends of the tropical curve yield the left ends of D as required. The balancing condition satisfied by (Γ, φ) implies the divergence condition at the white vertices. The black vertices correspond to marked points on elevators and are thus 2-valent and of divergence 0. Every elevator must be fixed by a marked point, or we would have a string. Thus D is bipartite. The genus of D equals the genus of the tropical curve. Thus D is a floor diagram of degree d and genus g. A plane tropical curve which is parametrized by a disconnected abstract tropical curve leads to a disconnected floor diagram. □

Proposition 11.2.7 *Let D be a floor diagram of genus g and degree d. There is a unique tropical curve of degree d and genus g passing through the points in the horizontal strip that yield D under the procedure described in Construction 11.2.6, and its multiplicity equals* mult(D).

Proof Consider the points in the horizontal strip. They are ordered according to the linear order of D. For each black vertex of D, draw an elevator of the right weight through the corresponding point. One connects these elevator germs to a tropical curve. Imagine first to start at the bottom of the page and draw an upward-pointing end in the vicinity of a point corresponding to a white vertex of D. By continuing to draw the end, it meets the lowest elevator (on the left or right). The balancing condition implies how to continue drawing. There is a unique way to complete the drawing to a floor connected to the elevators left and right of the point corresponding to the white vertex. Finally, shift this floor until it meets the point. In this way, one can see that there is precisely one tropical curve (Γ, φ) that yields D using Construction 11.2.6. By the genericity of the point conditions, (Γ, φ) is rigid and 3-valent.

Following Chap. 9, Definition 9.5.1, (Γ, φ) contributes a product of areas of triangles dual to (non-marked) vertices. Every compact edge e of D of weight $w(e)$ comes from a bounded edge e' of Γ of weight $w(e)$. Since e connects a white with a black vertex, e' is adjacent to precisely one non-marked 3-valent vertex V. Denote by e'' an edge in the floor which is adjacent to V. Every non-horizontal edge in a floor is of direction $(0, 1) + c \cdot (1, 0)$ for some c (by the balancing condition, the fact that the floor contains no cycles, and since one can connect every edge to an end of direction $(0, -1)$), and so the area of the triangle dual to V (formed by the duals of e' and e'') is $w(e)$.

Thus the product over all areas of triangles dual to non-marked vertices in the dual subdivision of (Γ, φ) equals the product of weights of the compact edges of D. □

11.3 Exercises

(1) Show that the numbers $N_{\text{trop}}^{\alpha,\beta}(d, g)$ do not depend on the point and end conditions.
(2) Think about a version of the Caporaso-Harris formula for irreducible tropical curves.
(3) Show that every floor of a tropical curve contains precisely one marked point.
(4) Compute $N_{\text{floor}}(3, 0)$.
(5) Can the definition of floor diagram be generalized to other polygons? Can it be generalized to count the numbers $N_{\text{trop}}^{\alpha,\beta}(d, g)$ showing up in the Caporaso-Harris formula [Blo12]?

(6) Following Exercise 3, find a floor diagram count that determines counts of tropical curves dual to a rectangle (corresponding to algebraic curves in $\mathbb{P}^1 \times \mathbb{P}^1$) [AB17]. Let us fix the weights of the left and right ends. Use this count to study the structure of these numbers: they are piecewise polynomial in the weights of the left and right ends. Find the positions of the walls dividing the regions of polynomiality. Compare with double Hurwitz numbers (see Part II).

Bibliography

[AB84] M. Atiyah, R. Bott, The moment map and equivariant cohomology. Topology **23**(1), 1–28 (1984)

[AB17] F. Ardila, E. Brugallé, The double Gromov-Witten invariants of Hirzebruch surfaces are piecewise polynomial. Int. Math. Res. Not. IMRN **2017**(2), 614–641 (2017)

[ABLdM11] A. Arroyo, E. Brugallé, L. López de Medrano, Recursive formulas for Welschinger invariants of the projective plane. Int. Math. Res. Not. IMRN **2011**(5), 1107–1134 (2011)

[AC14] D. Abramovich, Q. Chen, Stable logarithmic maps to Deligne-Faltings pairs II. Asian J. Math. **18**(3), 465–488 (2014)

[ACGS20a] D. Abramovich, Q. Chen, M. Gross, B. Siebert, Decomposition of degenerate Gromov-Witten invariants. Compos. Math. **156**(10), 2020–2075 (2020)

[ACGS20b] D. Abramovich, Q. Chen, M. Gross, B. Siebert, Punctured logarithmic maps. arXiv:2009.07720 (2020)

[ACP15] D. Abramovich, L. Caporaso, S. Payne, The tropicalization of the moduli space of curves. Ann. Sci. Éc. Norm. Supér. (4) **48**(4), 765–809 (2015). arXiv:1212.0373

[ACV01] D. Abramovich, A. Corti, A. Vistoli, Twisted bundles and admissible covers. Commun. Algebra **31**(8), 3547–3618 (2001)

[AK00] D. Abramovich, K. Karu, Weak semistable reduction in characteristic 0. Invent. Math. **139**(2), 241–273 (2000)

[ALT18] K. Adiprasito, G. Liu, M. Temkin, Semistable reduction in characteristic 0. arXiv:1810.03131 (2018)

[AW18] D. Abramovich, J. Wise, Birational invariance in logarithmic Gromov–Witten theory. Comput. Math. **154**(3), 595–620 (2018)

[BBM11] B. Bertrand, E. Brugallé, G. Mikhalkin, Tropical open Hurwitz numbers. Rend. Semin. Mat. Univ. Padova **125**, 157–171 (2011)

[BG16] F. Block, L. Göttsche, Refined curve counting with tropical geometry. Compos. Math. **152**(1), 115–151 (2016)

[BIMS15] E. Brugallé, I. Itenberg, G. Mikhalkin, K. Shaw, Brief introduction to tropical geometry, in *Proceedings of the Gökova Geometry-Topology Conference 2014*. Gökova Geometry/Topology Conference (GGT), Gökova (2015), pp. 1–75

[Blo11] F. Block, Computing node polynomials for plane curves. Math. Res. Lett. **18**(4), 621–643 (2011). arXiv:1006.0218

[Blo12] F. Block, Relative node polynomials for plane curves. J. Algebraic Combin. **36**(2), 279–308 (2012)

[BM08]   E. Brugallé, G. Mikhalkin, Floor decompositions of tropical curves: the planar case, in *Proceedings of the 15th Gökova Geometry-Topology Conference* (2008), pp. 64–90. arXiv:0812.3354

[Bou21] P. Bousseau, Refined floor diagrams from higher genera and lambda classes. Sel. Math. (N.S.) 27(3), Paper No. 43, 42 (2021)

[Bri97] M. Brion, Equivariant Chow groups for torus actions. Transform. Groups 2(3), 225–267 (1997)

[Bru15] E. Brugallé, Floor diagrams relative to a conic, and GW-W invariants of del Pezzo surfaces. Adv. Math. **279**, 438–500 (2015)

[BS14]   E. Brugallé, K. Shaw, A bit of tropical geometry. Amer. Math. Mon. **121**(7), 563–589 (2014)

[BS19]   L. Blechman, E. Shustin, Refined descendant invariants of toric surfaces. Discrete Comput. Geom. **62**(1), 180–208 (2019)

[BSSZ15] A. Buryak, S. Shadrin, L. Spitz, D. Zvonkine. Integrals of ψ-classes over double ramification cycles. Amer. J. Math. **137**(3), 699–737 (2015)

[Cap18] L. Caporaso, Recursive combinatorial aspects of compactified moduli spaces, in *Proceedings of the International Congress of Mathematicians—Rio de Janeiro 2018. Vol. II. Invited lectures* (World Scientific Publishing, Hackensack, 2018), pp. 635–652

[CGP21] M. Chan, S. Galatius, S. Payne, Tropical curves, graph complexes, and top weight cohomology of $\mathcal{M}_g$. J. Amer. Math. Soc. **34**(2), 565–594 (2021)

[CH98]   L. Caporaso, J. Harris, Counting plane curves of any genus. Invent. Math. **131**, 345–392 (1998)

[Che14] Q. Chen, Stable logarithmic maps to Deligne-Faltings pairs I. Ann. Math. **180**(2), 341–392 (2014)

[CJM10] R. Cavalieri, P. Johnson, H. Markwig, Tropical Hurwitz numbers. J. Algebraic Combin. **32**(2), 241–265 (2010)

[CJM11] R. Cavalieri, P. Johnson, H. Markwig, Wall crossings for double Hurwitz numbers. Adv. Math. **228**(4), 1894–1937 (2011)

[CJMR21] R. Cavalieri, P. Johnson, H. Markwig, D. Ranganathan, Counting curves on Hirzebruch surfaces: tropical geometry and the Fock space. Math. Proc. Camb. Philos. Soc. **171**(1), 165–205 (2021)

[CLS11] D.A. Cox, J.B. Little, H.K. Schenck, *Toric Varieties*, volume 124 of Graduate Studies in Mathematics (American Mathematical Society, Providence, 2011)

[CM14]   R. Cavalieri, S. Marcus, Geometric perspective on piecewise polynomiality of double Hurwitz numbers. Canad. Math. Bull. **57**(4), 749–764 (2014)

[CM16]   R. Cavalieri, E. Miles, *Riemann Surfaces and Algebraic Curves*, volume 87 of London Mathematical Society Student Texts (Cambridge University Press, Cambridge, 2016). A first course in Hurwitz theory

[CMR16] R. Cavalieri, H. Markwig, D. Ranganathan, Tropicalizing the space of admissible covers. Math. Ann. **364**(3–4), 1275–1313 (2016)

[CMR17] R. Cavalieri, H. Markwig, D. Ranganathan, Tropical compactification and the Gromov-Witten theory of $\mathbb{P}^1$. Selecta Math. (N.S.) **23**(2), 1027–1060 (2017)

[ELSV01] T. Ekedahl, S. Lando, M. Shapiro, A. Vainshtein, Hurwitz numbers and intersections on moduli spaces of curves. Invent. Math. **146**, 297–327 (2001)

[Fab99] C. Faber, Algorithms for computing intersection numbers on moduli spaces of curves, with an application to the class of the locus of Jacobians, in *New Trends in Algebraic Geometry (Warwick 1996)*, London Math. Soc. Lecture Note Ser., vol. 264 (1999), pp. 93–109

[FH91]   W. Fulton, J. Harris, *Representation Theory* (Springer, 1991)

[FM10] S. Fomin, G. Mikhalkin, Labeled floor diagrams for plane curves. J. Eur. Math. Soc. **12**(6), 1453–1496 (2010). arXiv:0906.3828

[FP97] W. Fulton, R. Pandharipande, Notes on stable maps and quantum cohomology, in *Algebraic Geometry—Santa Cruz 1995*, volume 62 of Proc. Sympos. Pure Math. (American Mathematical Society, Providence, 1997), pp. 45–96

[FP02] B. Fantechi, R. Pandharipande, Stable maps and branch divisors. Compos. Math. **130**(3), 345–364 (2002)

[FP18] G. Farkas, R. Pandharipande, The moduli space of twisted canonical divisors. J. Inst. Math. Jussieu **17**(3), 615–672 (2018)

[Ful93] W. Fulton, *Introduction to Toric Varieties* (Princeton University Press, 1993)

[Ful98] W. Fulton, *Intersection Theory*, 2nd edn. (Springer, 1998)

[Gat06] A. Gathmann, Tropical algebraic geometry. Jahresber. Deutsch. Math.-Verein. **108**(1), 3–32 (2006)

[GJ97] I.P. Goulden, D.M. Jackson, Transitive factorisations into transpositions and holomorphic mappings on the sphere. Proc. Amer. Math. Soc. **125**(1), 51–60 (1997)

[GJ99] I.P. Goulden, D.M. Jackson, A proof of a conjecture for the number of ramified coverings of the sphere by the torus. J. Combin. Theory Ser. A **88**(2), 246–258 (1999)

[GJV03] I. Goulden, D.M. Jackson, R. Vakil, Towards the geometry of double Hurwitz numbers. Preprint: math.AG/0309440v1 (2003)

[GJV06] I. Goulden, D. Jackson, R. Vakil, A short proof of the λ_g-conjecture without Gromov-Witten theory: Hurwitz theory and the moduli of curves. Preprint:mathAG/0604297 (2006)

[GKM09] A. Gathmann, M. Kerber, H. Markwig, Tropical fans and the moduli space of rational tropical curves. Compos. Math. **145**(1), 173–195 (2009). arXiv:0708.2268

[GM07] A. Gathmann, H. Markwig, The Caporaso–Harris formula and plane relative Gromov-Witten invariants in tropical geometry. Math. Ann. **338**(4), 845–868 (2007)

[GM08] A. Gathmann, H. Markwig, Kontsevich's formula and the WDVV equations in tropical geometry. Adv. Math. **217**, 537–560 (2008). arXiv:math.AG/0509628

[GM10] A. Gibney, D. Maclagan, Equations for Chow and Hilbert quotients. J. Algebra Number Theory **4**(7), 855–885 (2010). arXiv:0707.1801

[Gol21] C. Goldner, Counting tropical rational curves with cross-ratio constraints. Math. Z. **297**(1–2), 133–174 (2021)

[GP99] T. Graber, R. Pandharipande, Localization of virtual classes. Invent. Math. **135**(2), 487–518 (1999)

[GS13] M. Gross, B. Siebert, Logarithmic Gromov-Witten invariants. J. Amer. Math. Soc. **26**(2), 451–510 (2013)

[GS19] L. Göttsche, F. Schroeter, Refined broccoli invariants. J. Algebraic Geom. **28**(1), 1–41 (2019)

[Gub13] W. Gubler, A guide to tropicalizations, in *Algebraic and Combinatorial Aspects of Tropical Geometry*, vol. 589 (2013), pp. 125–189

[GV03] T. Graber, R. Vakil, Hodge integrals, Hurwitz numbers, and virtual localization. Compositio Math. **135**, 25–36 (2003)

[Has03] B. Hassett, Moduli spaces of weighted pointed stable curves. Adv. Math. **173**(2), 316–352 (2003)

[HKK+03] K. Hori, S. Katz, A. Klemm, R. Pandharipande, R. Thomas, C. Vafa, R. Vakil, E. Zaslow, *Mirror Symmetry*, volume 1 of Clay Mathematics Monographs (American Mathematical Society, Providence, 2003). With a preface by Vafa

[HM82] J. Harris, D. Mumford, On the Kodaira dimension of the moduli space of curves. Invent. Math. **67**, 23–88 (1982)

[HM98]   J. Harris, I. Morrison, *Moduli of Curves* (Springer, 1998)

[Hur91] A. Hurwitz, Ueber Riemann'sche Flächen mit gegebenen Verzweigungspunkten. Math. Ann. **39**(1), 1–60 (1891)

[IKS05] I. Itenberg, V. Kharlamov, E. Shustin,   Logarithmic asymptotics of the genus zero Gromov-Witten invariants of the blown up plane. Geom. Topol. **9**, 483–491 (2005)

[IMS09] I. Itenberg, G. Mikhalkin, E. Shustin,   *Tropical Algebraic Geometry*, 2nd edn., volume 35 of Oberwolfach Seminars (Birkhäuser Verlag, Basel, 2009)

[ISK09] I. Itenberg, E. Shustin, V. Kharlamov, A Caporaso-Harris type formula for Welschinger invariants of real toric Del Pezzo surfaces. Comment. Math. Helv. **84**, 87–126 (2009). arXiv:math.AG/0608549

[Joh14] P. Johnson, Equivariant GW theory of stacky curves. Commun. Math. Phys. **327**(2), 333–386 (2014)

[JPB20] A. Jaramillo Puentes, E. Brugallé,   Polynomiality properties of tropical refined invariants. Preprint, arXiv:2011.12668 (2020)

[JPT11] P. Johnson, R. Pandharipande, H.-H. Tseng,   Abelian Hurwitz-Hodge integrals. Michigan Math. J. **60**(1), 171–198 (2011)

[Kal17] N. Kalinin, Tropical approach to Nagata's conjecture in positive characteristic. Discrete Comput. Geom. **58**(1), 158–179 (2017)

[Kap93] M. Kapranov,   Chow quotients of Grassmannians. I, in *IM Gelfand Seminar*, vol. 16 (1993), pp. 29–110

[Koc01] J. Kock, Notes on psi classes. Notes. http://mat.uab.es/~kock/GW/notes/psi-notes.pdf (2001)

[Kon95] M. Kontsevich,   Enumeration of rational curves via torus actions,   in *The Moduli Space of Curves (Texel Island, 1994)*, volume 129 of Progr. Math. (Birkhäuser Boston, Boston, 1995), pp. 335–368

[KV07]   J. Kock, I. Vainsencher,   *An Invitation to Quantum Cohomology*, volume 249 of Progress in Mathematics (Birkhäuser Boston, Boston, 2007). Kontsevich's formula for rational plane curves

[Lew18] D. Lewanski, On ELSV-type formulae, Hurwitz numbers and topological recursion, in *Topological Recursion and Its Influence in Analysis, Geometry, and Topology*, volume 100 of Proc. Sympos. Pure Math. (Amer. Math. Soc., Providence, 2018), pp. 517–532

[Li01]   J. Li, Stable morphisms to singular schemes and relative stable morphisms. J. Diff. Geom. **57**(3), 509–578 (2001)

[Li02a] J. Li,   A degeneration formula of GW-invariants. J. Differ. Geom. **60**(2), 199–293 (2002)

[Li02b] J. Li,   A degeneration formula of GW-invariants. J. Differ. Geom. **60**(2), 199–293 (2002)

[LM00]   A. Losev, Y. Manin,   New moduli spaces of pointed curves and pencils of flat connections. Michigan Math. J. **48**, 443–472 (2000). Dedicated to William Fulton on the occasion of his 60th birthday

[LR01]   A.-M. Li, Y. Ruan, Symplectic surgery and Gromov-Witten invariants of Calabi-Yau 3-folds. Invent. Math. **145**(1), 151–218 (2001)

[Mar06] H. Markwig, The enumeration of plane tropical curves. PhD thesis, TU Kaiserslautern, 2006

[Mar10] T. Markwig, A field of generalised Puiseux series for tropical geometry. Rend. Semin. Mat. Univ. Politec. Torino **68**(1), 79–92 (2010)

[Mik05] G. Mikhalkin, Enumerative tropical geometry in $\mathbb{R}^2$. J. Amer. Math. Soc. **18**, 313–377 (2005)

[Mik07]   G. Mikhalkin, Moduli spaces of rational tropical curves, in *Proceedings of Gökova Geometry-Topology Conference GGT 2006* (2007), pp. 39–51. arXiv:0704.0839

[MMS18]   H. Markwig, T. Markwig, E. Shustin, Enumeration of complex and real surfaces via tropical geometry. Adv. Geom. **18**(1), 69–100 (2018)

[MR09]   H. Markwig, J. Rau, Tropical descendant Gromov-Witten invariants. Manuscripta Math. **129**(3), 293–335 (2009). arXiv:0809.1102

[MS15]   D. Maclagan, B. Sturmfels, *Introduction to Tropical Geometry*, volume 161 of Graduate Studies in Mathematics (American Mathematical Society, Providence, 2015)

[Mum83]   D. Mumford, Toward an enumerative geometry of the moduli space of curves. Arith. Geom. **II**(36), 271–326 (1983)

[MV14]   M. Melo, F. Viviani, The Picard group of the compactified universal Jacobian. Doc. Math. **19**, 457–507 (2014)

[NS06]   T. Nishinou, B. Siebert, Toric degenerations of toric varieties and tropical curves. Duke Math. J. **135**, 1–51 (2006)

[OP09]   A. Okounkov, R. Pandharipande, Gromov-Witten theory, Hurwitz numbers, and matrix models, in *Algebraic Geometry—Seattle 2005. Part 1*, volume 80 of Proc. Sympos. Pure Math. (American Mathematical Society, Providence, 2009), pp. 325–414

[Pan99]   R. Pandharipande, Hodge integrals and degenerate contributions. Commun. Math. Phys. **208**(2), 489–506 (1999)

[Pay06]   S. Payne, Equivariant Chow cohomology of toric varieties. Math. Res. Lett. **13**(1), 29–41 (2006)

[Ran17]   D. Ranganathan, Skeletons of stable maps I: rational curves in toric varieties. J. Lond. Math. Soc. (2) **95**(3), 804–832 (2017)

[Ran22]   D. Ranganathan, Logarithmic Gromov-Witten theory with expansions. Algebr. Geom. **9**(6), 714–761 (2022)

[RGST05]   J. Richter-Gebert, B. Sturmfels, T. Theobald, First steps in tropical geometry, in *Idempotent Mathematics and Mathematical Physics*, volume 377 of Contemp. Math. (American Mathematical Society, Providence, 2005), pp. 289–317

[Sau19]   A. Sauvaget, Cohomology classes of strata of differentials. Geom. Topol. **23**(3), 1085–1171 (2019)

[Shu06]   E. Shustin, A tropical calculation of the Welschinger invariants of real toric Del Pezzo surfaces. J. Algebr. Geom. **15**(2), 285–322 (2006). arXiv:mathAG/0406099

[SSV08]   S. Shadrin, M. Shapiro, A. Vainshtein, Chamber behavior of double Hurwitz numbers in genus 0. Adv. Math. **217**(1), 79–96 (2008)

[Tev07]   J. Tevelev, Compactifications of subvarieties of tori. Amer. J. Math. **129**(4), 1087–1104 (2007)

[Tyo17]   I. Tyomkin, Enumeration of rational curves with cross-ratio constraints. Adv. Math. **305**, 1356–1383 (2017)

[Uli21]   M. Ulirsch, A non-Archimedean analogue of Teichmüller space and its tropicalization. Selecta Math. (N.S.) **27**(3), Paper No. 39, 34 (2021)

[Vak08]   R. Vakil, The moduli space of curves and Gromov-Witten theory, in *Enumerative Invariants in Algebraic Geometry and String Theory*, volume 1947 of Lecture Notes in Math. (Springer, Berlin, 2008), pp. 143–198

[Wel05]   J.-Y. Welschinger, Invariants of real symplectic 4-manifolds and lower bounds in real enumerative geometry. Invent. Math. **162**(1), 195–234 (2005)

[Wil06]   H.S. Wilf, *generatingfunctionology*, 3rd edn. (A K Peters Ltd., Wellesley, 2006)

Printed in the USA
CPSIA information can be obtained
at www.ICGtesting.com
LVHW080303240124
769695LV00007B/514

9 783031 394003